Beekeeping for Gardeners

Beekeeping for Gardeners

The Complete Step-By-Step Guide to
Keeping Bees in Your Garden

Richard Rickitt

GREEN BOOKS

LONDON · OXFORD · NEW YORK · NEW DELHI · SYDNEY

GREEN BOOKS
Bloomsbury Publishing Plc
50 Bedford Square, London, WC1B 3DP, UK
29 Earlsfort Terrace, Dublin 2, Ireland

BLOOMSBURY, GREEN BOOKS and the Green Books logo
are trademarks of Bloomsbury Publishing Plc

First published in the United Kingdom 2024

A catalogue record for this book is available from the British Library.
Library of Congress Cataloguing-in-Publication data has been applied for.

ISBN: PB: 978-1-3994-0484-6
ePub: 978-1-3994-0483-9
ePDF: 978-1-3994-0485-3

2 4 6 8 10 9 7 5 3 1

Design by Susan McIntyre
Printed and bound in Turkey by Elma Basim

100%
From well-
managed forests
FSC® C164814

To find out more about our authors and books visit www.bloomsbury.com
and sign up for our newsletters

Contents

Introduction   6

PART ONE
Keeping honey bees   22

Introducing the honey bee   24
Becoming a beekeeper   36
How to get started   42
Keeping bees   58
Handling bees   68
How to manage swarming   80
Queen and apiary management   92
The rewards of beekeeping   106
Caring for bees in winter   128
Health and hygiene   134
The beekeeping year   158

PART THREE
Gardens for bees   208

A world outside your back door   210
What to plant for bees   218
Lawns, meadows and wild gardens   242
Pools and damp gardens   256
Hedges, shrubs and trees   264
Garden and farm crops   274
The best garden plants for bees   292

Further reading   296
Acknowledgements   298
Index   299

PART TWO
Bees in your garden   162

Bumblebees   164
Solitary bees   180

Introduction

The wonder of bees

It's often said that 'man's best friend' is the dog, but honey bees must surely run, or rather fly, a close second. Humans and honey bees have a shared history dating back hundreds of thousands of years. Indeed, bees were buzzing around long before we evolved, and even our most ancient ancestors must have known about the delicious treats that could be robbed from the nest of the honey bee. We certainly thought highly enough of bees to celebrate our relationship with them in some of our earliest artworks. Dated at about 5,000 years old, cave paintings found at Bicorp, eastern Spain, depict a person climbing a cliff and harvesting honey, as clouds of agitated bees buzz around them. Only slightly more recent are the numerous carvings of bees and hives on the walls of Egyptian temples and tombs. For millennia, honey bees have provided us with food by pollinating our crops, with light from their brightly burning wax, with various curatives from the natural medicine cabinet of their nest and, of course, with honey – the sweetest gift of all. Perhaps you are aware of this age-old bond between humans and bees, or had a grandparent who kept bees, or maybe you tried some local honey or saw something on TV that stirred your interest, but if you are reading this book, presumably you too have felt the curious and inexplicable urge that, throughout history, has made so many people think: 'I want to keep bees.'

It can come as a surprise to those new to the hobby that there is so much more to beekeeping than simply owning a beehive. It is a pastime – a craft – that is as multi-faceted as a piece of honeycomb. It can be approached from many angles and has something to offer almost anyone of any age. At a basic level it is about nurturing, cultivating and caring for honey bees. Like any kind of animal husbandry, there can be great satisfaction in doing your best for the creatures in your care and seeing them thrive. But don't expect this to

ABOVE Beekeeping relief at the tomb of Pabasa, near Luxor, Egypt, c.610 BCE.

OPPOSITE Beekeeping can be a delightful family activity that inspires young minds.

be anything like acquiring the average new pet. While beekeepers undoubtedly love their buzzing companions, no matter how you try to manage them your honey bees will remain resolutely independent, untameable and frequently unfathomable. It is perhaps the close interaction with such innately wild and mercurial creatures that for many people makes beekeeping so thrilling and fulfilling.

ABOVE Why not imbibe some homemade mead by the light of your own candles?

ABOVE There are rosettes, cups and prestige to be won at local and national honey shows.

If you crave solitude, beekeeping is a wonderfully restorative activity that allows you to lose yourself and your thoughts entirely within the world of the hive – it's just you, alone with perhaps 50,000 bees. But it can be a highly social activity too, whether you're sharing it with a partner or children, or with your friends. There is an active beekeeping community with local clubs and a range of regional and national events to attend, and even opportunities for bee-themed foreign holidays for those with understanding families.

Whether or not honey is your goal, healthy bees that are well cared for will normally produce a harvest,

BELOW Microscopy can bring a fascinating perspective to beekeeping.

and there can be pleasure in comparing different honeys from your own hives, and savouring the flavours and aromas of the concentrated nectar from a multitude of flowers. Processing honey is a skill in itself and there are many techniques to perfect. Honey can be used for cooking, baking and preserving, and for making beer and mead. You can enter it into local and national shows and sell it from the doorstep with the possibility of a modest profit. You will also have a harvest of wax, which can be used to make candles, cosmetics and food wraps as well as various forms of art.

Most beekeeping activity takes place in the warmest six months of the year, but in winter there is still plenty to do. If you enjoy DIY, you can make, adapt and even invent all kinds of beekeeping equipment. Or you can try your hand at microscopy; dissecting bees to reveal their fascinating anatomy, or studying pollen in all its wondrous, multi-faceted, minuscule forms.

Many beekeepers, previously only vaguely aware of the existence of other species of bees, develop a fascination with wild bumblebees and solitary bees. It is becoming increasingly obvious that these other species are suffering disproportionately from various environmental threats, and one aim of this book is to encourage new beekeepers to consider themselves guardians not only of the bees in their own hives, but also of the wild species that fly in the same skies and share the same floral resources.

There is a pleasing symbiosis between beekeeping and gardening. Many gardeners decide to try beekeeping after seeing bees in their garden, busily pollinating flowers and contributing to the improved production of fruit and vegetables. Conversely, many beekeepers previously without green fingers come to appreciate the importance of flowers to their bees, and develop an interest in growing bee-friendly plants – sometimes even planting extensive crops for honey production.

The days when you are likely to be out in your garden tend to coincide with the times when most beekeeping tasks need to be done, and so a weekend spent in the garden can encompass time spent with the bees. It is hard to imagine anything more satisfying than growing delicious produce carefully selected to bring you great pleasure, and seeing bees visiting them to collect pollen and nectar. What makes the experience even sweeter, quite literally, is knowing that your newly pollinated plants have contributed to both the health of local bee populations and to the jar of honey on your kitchen table.

Karl Von Frisch, who won the Nobel Prize for discovering the meaning of the honey bee's beguiling waggle dance, said that 'the bee's life is like a magic well: the more you draw from it, the more it fills with water'. Beekeeping can reward you mentally, physically, socially, culturally and even spiritually, deeply enriching your life and your environment in ways that I guarantee will continue to surprise and delight you for many years to come. It can indeed be an everlasting well from which you will never cease to draw.

The origins of a species

Bees are thought to have originated during the Cretaceous period, on the supercontinent of Gondwana, where flowering plants probably also first appeared. The connection is significant, of course, because bees need flowers, and flowers need bees. Starting some 140 million years ago, prehistoric insect-hunting wasps seem to have discovered that collecting and consuming pollen was easier than having to catch and kill for a living. These meat-eating wasps evolved into vegetarian bees, with bodies adapted to suck up sugary nectar, and to collect and transport

ABOVE Honey bees on a sunflower – a major agricultural crop.

dust-like pollen. Flowers evolved too, from plain and colourless to bright, scented, nectar-rich blossoms that flamboyantly advertised their wares. The earliest known fossil bee, found in a piece of Burmese amber, is thought to be almost 100 million years old. With the advent of gene mapping, we now know that within the taxonomic order Hymenoptera (which also includes ants, wasps and sawflies) there are seven families of bees divided into some 700 genera. Between them, these genera contain approximately 20,000 known species – with probably many more still to be discovered. There are more kinds of bee in the world than all the species of birds and mammals combined, ranging from the magnificent Wallace's giant bee (*Megachile pluto*) with a 6cm wingspan, to the 2mm-long *Perdita minima*, so tiny it doesn't even warrant a common name. Despite their ancient ancestors' conversion to a plant-based diet, there are three species of meat-eating 'vulture' bees.

Honey bees are in the family Apidae and the genus *Apis*, comprising eight species. They are social bees that have a queen, live in large colonies and make honey and wax. Most are confined to South East Asia and include the giant honey bee (*Apis dorsata*), the dwarf red honey bee (*Apis florea*) and the eastern honey bee (*Apis cerana*), whose small colonies can be kept in hives and managed for pollination and honey.

ABOVE A small hive for small bees – the eastern honey bee (*Apis cerana*) at an apiary in India.

RIGHT The giant honey bee (*Apis dorsata*), from Asia.

OPPOSITE PAGE A male hairy-footed flower bee (*Anthophora plumipes*), a type of solitary bee, approaching cherry blossom.

However, throughout the world most bees kept in hives are *Apis mellifera*, the European or western honey bee. The origin of these bees has been much debated, but DNA meta-analysis now suggests that their ancestors evolved in western Asia about seven million years ago. Moving via Asia Minor, one branch radiated into eastern Europe while another spread into northern Africa and then progressed into southern Europe via the Iberian Peninsula. These bees evolved into the various regional subspecies known today, some of which have been exported all over the world to be kept in hives for crop pollination and the production of wax and honey.

As well as honey bees, there are various species of bumblebee and solitary bee. Bumblebees are closely related to honey bees and are in the same family, Apidae, although they are in the genus is *Bombus*. They share some features with honey bees, such as hairy bodies and legs adapted to collect and carry pollen. They live in modest-sized colonies and make nests

using wax, but they do not make or store honey. Some solitary bees are also in the Apidae family, including the *Anthophora* flower bees and the *Nomada* cuckoo bees. The remaining 200 or so UK solitary bee species are in five other families, the most numerous being members of the Colletidae, Andrenidae and Halictidae. Most solitary bees do not live in any kind of social group but live and work alone, gathering food, building nests and laying eggs in a fascinating range of lifestyles which are covered in more detail on pages 181–207.

The importance of bees

Albert Einstein is supposed to have said that without bees, humans would survive only four more years. There is no evidence that the Nobel prize-winning physicist ever said this, but if he did it's fair to say that he probably hadn't done his homework. Bees are of course important pollinators, but so too are flies, wasps, ants, midges, butterflies, moths, beetles, birds, bats and even some reptiles.

Bees, though, are among the most important pollinators. Their hairy bodies are perfectly designed for picking up and carrying pollen, and their habit of flying back and forth between flowers of the same species – a trait known as flower fidelity – means that they are highly effective. They are also very reliable; thousands of species exist in almost every environment on earth, meaning that there is nearly always a bee around to pollinate a flower – and sometimes there is even a bee that has evolved to pollinate that specific flower.

It is through the food we eat that most of us feel the direct benefit of pollinators. Estimates vary, but a generally accepted figure is that one mouthful in every three can be directly attributed to pollinators visiting plants. Another metric is that about 75 per cent of the world's major crop species rely on pollinator visits, meaning that although important staples like maize, rice and wheat can be produced without pollination, most of the highly nutritious plant foods – and generally the more enjoyable ones – rely on pollinator visits. These include fruit, vegetables, nuts, most coffee and even chocolate.

Many foods that don't require direct pollination to produce a crop still require pollination within their life cycle. For example, a carrot seed can grow into an edible root without pollination, but the seed itself will have come from a carrot flower, which must have been visited by a pollinator. Many crops are self-fertile, yet will produce a larger or better-quality crop if visited by pollinators. Cotton plants, for instance, can be up to 60 per cent more productive and grow a more durable and valuable fibre if visited by pollinating insects, most often by managed honey bees. Meat and dairy products are not included in the one mouthful in three calculations because farm animals don't have to be pollinated, yet many of the plants on which they feed depend to some degree on pollination.

In 2019, worldwide agricultural production resulting from pollinator visits was calculated by the United Nations (UN) to be worth about $600 billion. An incidental statistic is the value of the world honey market, worth a further $8 billion – plus tens of thousands of beekeeper livelihoods.

The UN estimates that some 800 million people practise urban agriculture in gardens, allotments and even window boxes. In such cases the monetary value of pollination is impossible to determine, except to say that millions of mouths are fed daily because of the services freely provided by pollinators. Undoubtedly, without pollinators the life chances of many of the world's poorest people would be severely affected;

a 2022 Harvard University report estimated that annually some 500,000 early deaths can now be linked to lower yields of the most nutritious crops as a result of pollinator declines.

But perhaps more than food and financial considerations, we must be grateful to bees and other pollinators for the role they play in sustaining the richly abundant ecosystem on which we rely for our health, wellbeing and very existence. The flower-visiting activities of pollinators are at the very heart of the web of plants, animals, birds, insects, fungi and microbes that continue, in however degraded and fragile a state, to support human life on Earth.

Insect Armageddon?

Since you are reading this book, you are doubtless already very aware that bees and many other insects are threatened in a way that could prove catastrophic. But it doesn't take the almost constant stream of alarming media stories to remind most of us that there is a problem; we can see it with our own eyes – or rather, we cannot. Not that long ago, summertime road trips in the UK usually resulted in a car covered with squashed insects. But when was the last time you saw anything go splat on your windscreen? And, until relatively recently, every street lamp was surrounded on summer nights by a halo of fluttering insects and gratefully swooping bats. Perhaps to the relief of some, that rarely happens now.

Beyond the anecdotal evidence, hundreds of academic studies frequently highlight significant declines in insect populations across the world. Probably the best known of these was undertaken by Germany's Krefeld Society, which trapped insects across 63 sites between 1989 and 2016. The analysis showed an 82 per cent drop in insect biomass during the 17-year study. Worryingly, the tests were conducted in nature reserves, places where you might expect insects to be least endangered. The trend has been confirmed by other peer-reviewed reports that are consistent, undeniable and international – the world is losing its insects at a staggering rate.

Habitat loss is a huge driver of declines. Continuing tropical deforestation is horrifying and deserves the focus of our attention, but environmental destruction continues all around us in myriad ways, big and small, that eat away at the foundations of the natural world: the building of roads, housing estates, industrial buildings and shopping centres; the cutting down of street trees; the paving over of front gardens; and the laying of plastic lawns. Every square metre made uninhabitable diminishes the opportunities for plant and insect life to survive.

BELOW Oilseed rape as far as the eye can see.

ABOVE Corn fields were once bejeweled with annual flowers such as poppies and corn marigolds.

The damage done to our countryside since the Second World War is staggering. The UK has lost about half of its hedgerows (about 200,000 km), and those that remain are often little more than living fences, flailed annually into spindly skeletons where barely a bird or a bee can nest or find sustenance. Many irreplaceable ancient woodlands have been bulldozed or replaced by all-but-sterile conifer plantations. Some 97 per cent of our wildflower meadows have been lost, their biodiverse tapestry of grasses and flowers – once managed on a cycle that benefitted both humans and wildlife – replaced by green deserts of rye grass, supercharged with nitrogen fertiliser, sprayed to kill flowering plants, and harvested several times a year to feed livestock that live indoors under fluorescent lighting.

Driven by successive governments to produce ever more and cheaper food, farmers rely on vast quantities of agrochemicals to achieve a worthwhile harvest from monocultured crops growing in almost sterile soils. In 1962, Rachel Carson's ground-breaking book *Silent Spring* revealed the insidious effects on wildlife and the environment of DDT insecticide, yet the substance wasn't banned in the UK until 1984, and was used worldwide for a further 20 years. Organophosphate and pyrethroid insecticides continue to be sprayed, drifting into hedgerows, ponds and surrounding areas to poison all lifeforms, whether they threatened crops or not.

Beginning in the 1990s, an apparent improvement in the form of neonicotinoid insecticides (neonics) began to be used. Rather than spraying crops repeatedly, a powdered coating on each planted seed ensured that the poison was absorbed by the growing plant. Initially this seemed to be a welcome alternative because it cut down on fuel use and the dangers of drifting spray. It also meant that the poison was built into plants systemically, affecting only the creatures that fed on them. It wasn't long before problems became evident; bees and other insects continued disappearing in ever larger numbers, along with the creatures that ate them. It was discovered that, although minute quantities of neonics can kill a bee (perhaps only about four billionths of a gram in the case of a honey bee), even more minuscule amounts consumed via nectar and pollen can have numerous sublethal effects. Among other things, these chemicals are neurotoxins; bees that suck up neonic-laced nectar become disoriented and forget their way home, flying around hopelessly until they die. More sublethal effects are being discovered all the time; neonics affect bees' ability to learn, communicate, metabolise, lay eggs, properly raise young and to fend off a range of diseases. They have equally damaging effects on butterflies, moths, beetles and other insects. And whereas it was thought that the poison is safely locked up inside only those plants whose seeds have been coated, actually a large proportion of the powder washes into the ground, where it remains active for a long time. Residues can find their way into watercourses, trees, hedgerow plants and even 'bee-friendly' wildflower strips, turning them into toxic nectar bars.

In 2013 the EU banned the outdoor use of several neonicotinoids, and others followed. One neonicotinoid, acetamiprid, is still permitted for use in the UK in fruit orchards, cereal crops and potted plants, and in a range of widely available chemical products sold for use by amateur gardeners. Shockingly, some 300,000 tons of neonicotinoids are manufactured annually in the EU and the UK, although their use is banned here; instead, they are exported for use by the rest of the world. To put that into perspective, one teaspoon of the neonicotinoid thiamethoxam has been calculated to be sufficient to kill 1.25 billion bees.

As well as insecticides, other agrochemicals are increasingly being shown to adversely affect the ecosystem. These include not only fungicides, but also herbicides such as glyphosate – a staggering one million tons of which is used globally each year. We are beginning to realise that in some cases it is the other ingredients that these chemicals are mixed with to produce various proprietary products that can make them particularly harmful. Because these other ingredients are not 'active' they are not as well regulated, and their effects are less understood. Our knowledge of what happens when cocktails of these products accumulate in soil and water – and ultimately our own bodies – is almost non-existent.

In addition to agrochemicals administered on an industrial scale, what other perils might face our bees and other insects? It is possible that diesel particulates and various airborne industrial pollutants might be causing problems, and perhaps airborne microplastics are clogging bees' minute breathing tubes – millions of particles can be generated every time a tumble dryer is used, for example. And, of course, the climate crisis looms large over everything. It's not yet clear exactly how the increasingly apparent symptoms of climate change will affect many species, although altered seasonal patterns are already disrupting the synchronisation between some bees and the plants on which they depend. Bumblebees, which prefer a cooler climate, are among the more obvious likely losers.

In the UK, the volume of insects has diminished significantly, but so far relatively few have gone extinct.

The good news is that, in the right conditions, insects multiply rapidly and could bounce back if given a chance. There might only be a short time left for this to happen, though, so we need to start supporting our environment now, before it is no longer capable of supporting us.

How are the bees?

If you become a beekeeper, prepare to be frequently asked, 'How are the bees?' The answer, as already suggested, is that bees are in trouble. But the situation is somewhat nuanced; some bees are more endangered than others and, perhaps surprisingly, a few are currently of very little concern indeed – and they may not be the ones you might expect.

Since 1900, the UK has lost two species of bumblebee, and a further eight species are considered in serious danger. More than 20 species of solitary bee are now extinct in the UK, and about a dozen are endangered. The general picture is that the numbers of most species of bees have declined by up to a third, and that they now tend to be found in fewer, smaller areas. For example, the once quite common shrill carder bee (*Bombus sylvarum*) is now confined to just a few small locations, and even in some of those

ABOVE Once common, the shrill carder bumblebee is now one of our rarest bees.

it hasn't been seen recently. Some species remain stable, sometimes increasing their range though not necessarily their numbers. Only two species have shown significant increases, and these are both recent arrivals from mainland Europe; the tree bumblebee (*Bombus hypnorum*) and the ivy bee (*Collettes hederae*) have each established themselves very successfully in the south and rapidly extended their range northwards as far as Scotland. But while some species of bees are very widespread, they may have quite small populations that are very localised, sometimes being found in only a few restricted locations in a handful of counties. This makes their existence very fragile; an entire local population could be threatened by a single building development or a change in agricultural practices on just one farm.

When most people express concern about bees, what they usually mean is honey bees. This worry stems largely from several quite serious events that generated worldwide headlines and predictions of imminent catastrophe. In 1992 the parasitic varroa mite, originating in Asia, was first discovered in the UK, having probably arrived with some unwisely imported bees. Varroa soon caused the death of many colonies while beekeepers struggled to find ways to tackle what soon became a widely reported problem. At about the same time, huge numbers of honey bees died in several mainland European countries. These die-offs were thought to be linked to the use of neonicotinoid insecticides, and sparked large protests which were reported in the media worldwide. The number of honey bee colonies in the USA had been on the decline since the 1960s, but in the mid-2000s, US beekeepers started to report the death of monumental quantities of bees. Often, previously healthy-seeming colonies were disappearing almost overnight; tens of thousands of bees were vanishing and leaving hives empty except for some young and the queen. Some commercial beekeepers lost tens of thousands of colonies in this way. At the time, these events were a mystery. Today, it's thought that a range of causes, including varroa, diseases, pesticides, habitat loss and stress from insensitive management, contributed to what became known as colony collapse disorder (CCD).

In some countries the numbers of beekeepers and honey bee colonies fell dramatically from the 1990s. The British Beekeepers Association (BBKA) had 15,000 members in 1990, falling to 9,000 in 2001. Stories about the decline of beekeeping, threats to honey bees and the coming pollinator apocalypse appeared regularly in the media, and this had the effect of stirring some people into action. Suddenly, everyone wanted to save the bees. Gradually, beekeeper numbers began to climb again, until in 2022 the BBKA had 27,000 members.

Today, honey bees face many of the same threats as bumblebees and solitary bees, but generally beekeepers can keep their colonies healthy through various management techniques, something that cannot be done for wild species. Estimates vary, but it's thought that in the mid-1950s there were in the region of 250,000 managed colonies of honey bees in the UK. Government statistics for 2021 estimated a figure of 272,000 colonies, suggesting that, after earlier declines, honey bee numbers in the UK are buoyant. Even so, research undertaken at the Rothamsted Research (the UK's leading agricultural research institute) suggest these numbers could be historically quite low, and that there may have been as many as one million managed hives in the UK before the First World War, and possibly even more in the 1930s. Worldwide, it is thought that honey bee populations are now larger than ever before.

We are bombarded with media messages to save the bee, and encouraged to buy bee-friendly products – from wildflower seeds to food and cosmetics. Rarely do campaigns or products specify which bees need saving, but if an image is used it's almost always that of a honey bee. Honey bees are extraordinary creatures, rightly admired for their role in pollinating our food plants and in making delicious honey. But although they are vulnerable and in need of much more research into their needs, honey bees are not currently endangered. If your desire is to help save the bees, become a beekeeper by all means, but devote some of your effort to the often overlooked wild bumblebees and solitary bees, which deserve just as much attention as their hive-dwelling cousins.

The wrong kind of bees?

One result of the recent rise in popularity of beekeeping has been the claims by some ecologists and entomologists that honey bees are outcompeting solitary bees and bumblebees, exacerbating their decline. In major cities at least, they have a point; anywhere with more concrete and glass than trees and flowers will be a challenging environment for any kind

ABOVE Hives on the roof of a central London office block. Urban beekeeping has seen a huge rise in popularity.

of bees, so introducing millions of highly efficient honey bees is bound to be problematic. Indeed, bees in central London are already recorded as exhibiting signs of stress associated with poor nutrition. Nevertheless, the arguments are more complicated than often presented, and the negative portrayal by some of honey bees and beekeeping is unhelpful, and sometimes disingenuous. Part of the reason seems to stem from the frustration that so many sources, particularly in the media but also some beekeepers, continually and incorrectly claim that keeping honey bees is helping to save pollinators. One of the counter-claims commonly made in reaction to this is that the honey bee is a non-native, domesticated species.

Addressing whether honey bees are native to the UK, the following argument can be made: the honey bee (*Apis mellifera*) and various species of solitary bees and bumblebees would have spread north and eastwards from mainland Europe as the ice receded after the last glacial maximum. About 6,000 years ago, rising sea levels separated what would become the British Isles from the European mainland, isolating those species of bees already established there. The accepted definition of native in this context is any species established in Britain before it was surrounded by water and which was not subsequently introduced by humans. Therefore, honey bees are probably as native to Britain as are all

other long-resident species of bee. Worth noting here is that although many species of solitary bees and bumblebees are present and considered native in both the UK and mainland Europe, only the honey bee has been labelled by some as a UK non-native.

The issue of domestication is more nuanced and offers an interesting insight into humanity's long-established relationship with honey bees. Honey was the only reliable and affordable source of sugar in Europe until about the seventeenth century, when the system of using slaves to grow and harvest cane sugar was so hideously perfected. At first, all honey bees would have been wild-living and their honey and wax harvested from nests, predominantly in hollow trees. Later, we learned to house wild swarms in straw baskets called skeps. From the mid-nineteenth century, movable-frame wooden hives gradually came into use, allowing bees to be permanently housed and managed in ways somewhat comparable to those for other livestock. Only in the last 150 years have attempts been made to hybridise honey bees, crossing the best-performing strains to produce more productive, healthier bees. Some of the subspecies used in UK breeding programmes have been imported, such as *A. m. ligustica* from Italy and *A. m. carnica* from central Europe. Although still the same species of honey bee, these have evolved over thousands of years into subspecies, with small, mostly behavioural differences that make them best adapted to cope with the varied habitats, climates and flora of their region.

In the UK today, many commercial beekeepers use home-bred or imported hybrid honey bees. The majority of hobby beekeepers, although sometimes starting out with purchased hybrids, tend to have what are known as local bees. After successive generations, these have mixed with the wider local population to produce bees with a broad genetic mix. Because it tends to be the more successful colonies that are able or allowed to reproduce, the result can be bees particularly suited to local conditions; these are known as ecotypes. The fly – or queen bee – in the ointment is that if another local beekeeper imports bees from elsewhere, this can introduce a new and potentially destabilising genetic influence to an established population. There is now a widespread feeling within UK hobby beekeeping that queen bees should not be imported, and that keeping local bees is more sustainable. There is also a movement to reintroduce or increase the prevalence of *A. m. mellifera*, the native dark honey bee subspecies which was found in the UK and throughout north-western Europe before various other subspecies were introduced.

BELOW The colouration of these two worker honey bees hints at their differing genetic heritage.

Honey bees could be said to be domesticated if your definition of the term encompasses all creatures that have been selectively bred to fulfil the needs of humans. But whether they are pure *A. m. mellifera* or have a mix of genes from various subspecies, honey bees today don't look or behave significantly differently to their ancient ancestors. Another definition of domestication might be whether livestock are kept in an enclosed or artificial environment. It is the honey bee's obliging habit of living in almost any suitable cavity we provide them with that makes it possible for us to efficiently manage them and harvest their honey. And while it is true that most honey bee colonies are housed in specially constructed hives, many are also unmanaged and free-living in chimneys, wall cavities and hollow trees. Indeed, recent research suggests that free-living honey bee colonies nesting in trees in Britain and Ireland are far more common than was previously thought. Finally, apart from the highly specialised and rarely used practice of instrumental insemination, honey bees cannot mate in captivity, and therefore do not meet a key requirement of domestication.

Responsible beekeeping

Rather than dwelling on whether honey bees are native or domesticated, the important question from a wider conservation perspective is whether there are too many of them. Although the number of managed honey bee colonies in the UK is probably far fewer than 100 years ago, since then there has been a huge loss of habitat as well as heavy use of agrochemicals. And whereas beekeeping was once predominantly a rural and suburban activity, it is now a common hobby in urban and inner-city areas. So, although the number of honey bees today might be fewer than in the past, the ability of some environments to sustain them without impacting other species is perhaps reduced. Hard evidence one way or the other on the effects of beekeeping on other bees is scarce, and much research is ongoing, but when it comes to the best places to keep honey bees it is fair to say that there are varying scenarios for different locations.

Inner cities and urban areas with few green spaces or flowering plants may be unable to adequately support both wild bees and honey bee colonies, even if there are only small numbers of the latter.

Suburban areas often have extensive parks and gardens as well as allotments, brownfield sites and road and railway embankments. Many species of wild bee do well in such places, and these areas can support honey bee colonies in moderate numbers.

Small and semi-rural towns can have a mix of parks and gardens, many street trees, and be surrounded by wildlife reserves and agricultural land. Such places can have significant populations of wild bee species and are also among the most sustainable places to keep honey bees.

Sparsely populated rural areas with small, mixed farms and plenty of hedges, woodlands, river banks and small towns and villages with gardens offer a wide variety of nesting habitats and forage for wild bees, and can support reasonable numbers of honey bees.

Intensively farmed land is one of the worst environments for wildlife of all types. Large-scale monocultures can offer significant foraging opportunities for bees of many kinds for short periods, after which there are few flowers, hedges, trees or gardens to offer habitat or resources. Honey bees are best moved to flowering crops, whose

pollination and productivity they can positively influence, and taken away afterwards.

Sensitive conservation areas where some species of wild bee are rare or endangered should not have additional competition imposed by the nearby siting of anything more than very small numbers of sensitively managed honey bee colonies.

If by now you are feeling gloomy about the prospects and purpose of becoming a beekeeper, please don't be. My intention here is not to put you off beekeeping, or to make you think it is impossible to keep bees where you live, but to highlight that, unlike keeping many other kinds of pet or livestock, beekeeping is not an activity that exists in isolation. Free-roaming honey bees are part of the landscape in which they live. They have a relationship with the natural world that is overwhelmingly positive, but with potential to have some negative impacts, within an environment stretching several kilometres in all directions from their hive – about as far as honey bees typically fly. Certainly, the number of colonies within a few mostly inner-city locations is too high, but hive density in the UK is generally quite low, and in many places apiaries are few and far between.

ABOVE The Chichester apiary of Mr J Daniels in 1902. Few sites today could support so many colonies.

Generally, it's fair to say that beekeepers have traditionally had a somewhat blinkered view of honey bees as being the most important species of bee – perhaps even the only one worth caring about. Increasingly it is recognised that this is not the case, that other species of bees and pollinating insects are equally important and that their populations are fragile. In this book, I encourage new and established beekeepers to consider how the honey bees in their care fit into the local environment, and what can be done to ensure that all pollinators have an equal chance to survive and thrive in an increasingly challenging world. For both new and existing beekeepers, I suggest the following guidelines to help you to make the most of your hobby while achieving these goals.

- Keep only enough honey bees for your own enjoyment or the production of modest quantities of honey.
- Consider your local environment and landscape. Can it sustain your bees without affecting wild bee populations?
- Join your local beekeeping association. Ask them about beekeeper and colony numbers in your area, and find out about the association's related policies.
- Complete a beginner's course and gain experience handling bees before you get your first colony.
- Learn about honey bee diseases and strive to keep healthy honey bees that won't spread disease.
- Learn about swarming and how to keep it to a minimum.
- Buy only UK-bred, preferably local bees and queens.

- Learn about the local wild species of bumblebees and solitary bees. Ask local wildlife trusts if there are any sensitive areas where honey bees are best not kept, or any rare or endangered species that you can help to support.
- Plant as many bee-useful plants and trees as you can in your own garden or property, and encourage others to do the same.
- Keep bees as sustainably as you can, and garden organically.
- Ask your local council and businesses about their pollinator strategies. Encourage them to plant more for pollinators and to adopt organic land management policies.
- Become a bee ambassador. Use your knowledge and enthusiasm to inform others about the differences between honey bees, bumblebees and solitary bees, and what they can do to help all of them.

Learning about beekeeping

This book will explain the basic principles involved in keeping honey bees, but it won't make you a knowledgeable, fully competent beekeeper. That will take time, more reading, guidance from local beekeepers and, most importantly, the kind of experience that can only come with directly managing bees through many seasons, each of which is never the same. Many other books are available, some of them covering in great detail specific aspects of honey bee biology, behaviour and management. The UK monthly magazine *BeeCraft* has for over a century been a fantastic source of information and practical, seasonal guidance for beekeepers of all levels. There are thousands of online beekeeping videos too, but these should be treated with some caution. Many are made by experienced and trustworthy beekeepers, but others are somewhat inaccurate and misleading. As a beginner, it's difficult to tell them apart. As a rule, avoid videos and books from outside your own country or region as the seasons, plants, equipment and regulations may be different.

Getting help and advice from experienced beekeepers is invaluable. The best way to do this is to join your nearest beekeeping association. Most hold monthly meetings with interesting talks and offer a chance to meet other, experienced local beekeepers. Usually, beginner classes are offered, and this is a must if you are serious about keeping bees. Often you will be paired with an experienced beekeeping mentor for the first year or two of your exciting but sometimes baffling new hobby. About the most useful thing you can do as a beekeeper at any level is to see more bees. The more hives you see opened in different places by different beekeepers, the more you will come to understand the nature of honey bees and the craft of keeping them.

A note about seasons

Beekeeping is a seasonal hobby, and most situations and tasks described in this book will occur annually and in roughly the same order. However, exact timings will differ from year to year, particularly now that climate change is making traditional seasonality somewhat elastic. Geography plays its part too – what Cornish bees do in June might not happen for another month in Aberdeenshire. In this book, I avoid ascribing various

beekeeping activities to specific months and refer instead to seasons, each subdivided. For example, the early part of the year is divided into early spring, mid-spring and late spring. The exact timing could change from year to year and depending on location. On pages 158–161 there is a summary of a typical beekeeping year, broken down into seasons, with suggestions as to when those seasons might occur in the south of the UK. If you live elsewhere, adjust your timings accordingly. When writing about plants, I have referred to traditionally accepted flowering times, usually by month. Most keen gardeners know whether their plot is typically a little early or late, and are close observers of the subtle differences in timing from one year to the next. This is something which, in gardening and beekeeping alike, makes every season slightly different, every year a little challenging and the whole process continually fascinating and rewarding.

ABOVE Before you get any bees, attend a beekeeping course.

PART ONE

Keeping honey bees

The western honey bee, *Apis mellifera*, is a superb example of evolutionary perfection. Every body part has developed to perform a specific function with absolute efficiency. The cleverest human engineers could hardly design a machine better suited to its job. It's no wonder Charles Darwin found it hard to explain how a creature so eminently fit for purpose could be the result of natural selection rather than divine intervention.

Introducing the honey bee

Honey bee anatomy

At first glance, bees appear anatomically quite simple. Their bodies are divided into three clearly defined sections – head, thorax and abdomen. Each of these has a tough outer shell, the exoskeleton, containing the soft inner parts. Bees also have six legs, four wings and two antennae. But looking closer reveals fascinatingly complex, sometimes microscopic, features, and explains much about how bees live, work, communicate and reproduce. Understanding a little about honey bee anatomy will help you not only to be a better beekeeper, but also to appreciate the privilege of working so closely with one of nature's most marvellous creations. There follows a very basic introduction to the key features of the honey bee, but as your relationship with these creatures develops, you will discover that there is much more to see, understand and be bedazzled by.

RIGHT TOP The anatomy of a worker honey bee laid bare by the scanning electron microscope (SEM).

RIGHT Worker honey bee on a flower.

OPPOSITE PAGE Anatomical acrobatics: a honey bee worker uses the pollen brush on her leg to preen pollen from her wing. After dampening with nectar, it is stored on her hind leg.

PREVIOUS PAGES A worker honey bee uses her proboscis to probe the nectaries of wild clematis.

The head

The most noticeable features of a honey bee's head are its two large eyes. They are compound eyes made up of thousands of hexagonal lenses, each gathering and processing light to produce an image – similar to a digital camera producing images using pixels. A worker bee's eye has around 6,000 lenses, whereas a drone's eye has about 10,000. But bees don't have only two eyes; on their forehead are three more. These tiny eyes are called ocelli, or simple eyes. They cannot perceive images but receive and interpret wavelengths of light, which assist with orientation and navigation.

Also prominent on a bee's head are the two horn-like antennae, used for sampling the world around them. These are covered with hundreds of thousands of sensilla – tiny receptors that detect and measure stimuli including vibrations, temperature, humidity, carbon dioxide and communication pheromones emitted by other members of the colony.

At the bottom of the head are the beak-like jaws, called mandibles, which are used for biting, grooming and sculpting wax. Inside the mandibles are two glands. When a worker bee is young and functioning as a nurse bee, these mandibular glands produce a milky liquid which is one of the components of brood food, the diet of honey bee larvae. The liquid secreted by the glands

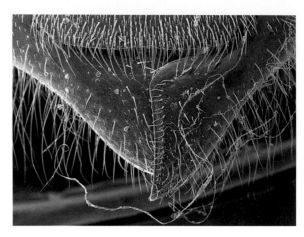

ABOVE Worker mandibles.

drains down a channel in the mandibles to be delivered one drop at a time to the young. When bees mature, these glands switch to producing a pheromone which is used as part of the colony's defence mechanisms.

Tucked beneath the head are the components of the proboscis. These unfold and zip together to produce a straw-like tube for sucking up water and nectar or for passing food between bees, a process called trophallaxis. A long tongue with a spongy end for lapping up liquids can extend down inside the proboscis. Nectar sucked

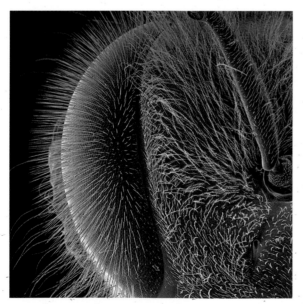

ABOVE SEM image of worker's face showing sensilla on antenna and feathery hairs of face and eyes.

ABOVE The extravagantly hairy head of a honey bee – they have some three million hairs in all.

from flowers travels up the proboscis and through the head where it passes a duct connected to the hypopharyngeal gland – an organ in the top of the head that produces enzymes which help to convert nectar into honey. The enzymes are added to the incoming nectar, and various sugar-changing chemical reactions are set in motion. In young bees, the hypopharyngeal gland serves another purpose – making a clear liquid which is added to the secretions of the mandibular glands to produce brood food.

The thorax

The middle section of the bee is called the thorax, most of which is made up of muscles that run horizontally or vertically. These can be stretched and relaxed hundreds of times a second, distorting the thorax from a sphere to a rugby-ball shape and back again. This flexing makes the wings, attached to the top of the thorax, flap up and down in a figure-of-eight pattern, enabling flight. There are four wings – two large forewings and two smaller hindwings. When resting they are separated and lie angled backwards over the bee's back. For flight, they are brought forward and zipped together using tiny hooks called hamuli – a bit like Velcro. The hard-working muscles require a lot of oxygen, which enters through valves in the sides of the thorax called spiracles. The thorax has two pairs of large spiracles, and there are six pairs of smaller spiracles along the sides of the abdomen. The tubes that lead from the spiracles, called tracheae, divide and become increasingly fine as they transport oxygen throughout the body. Running through the centre of the thorax is the oesophagus, a tube that transports nectar, pollen and water from the head to the abdomen.

The abdomen

The largest segment of a bee, the abdomen, contains most of the major organs. The outer case of the abdomen is made up of overlapping shell-like plates; those on top of the bee are called tergites, while on the underside are the sternites. These allow the abdomen to flex, a bit like an armadillo's body. On the underside of a worker's abdomen, between the overlapping sternites, are four pairs of wax glands. These secrete liquid wax onto the smooth surfaces of the sternites, known as wax mirrors, hardening to form tiny glass-like flakes of beeswax which are harvested, chewed to plasticise them, and used to build comb.

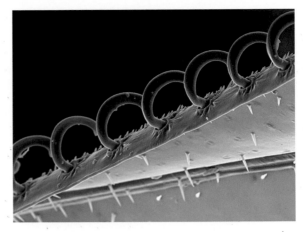

ABOVE Tiny hooks called hamuli zip the wings together.

A worker can bend the end of its abdomen to open a gap between two tergites, revealing a membrane with a groove or canal in it. The nearby Nasonov gland secretes a highly volatile oily pheromone into the canal, from where – with the aid of fanning wings – it is evaporated and wafted outwards. Nasonov pheromone works as a kind of homing signal; if a colony is disturbed, or on the move when swarming, guide bees will emit this lemon-scented chemical so that other bees can follow them.

The segments at the very end of a worker bee's abdomen open up to reveal the sting, an elegantly engineered mechanism which, when pushed into an enemy, will drill itself further down into their flesh while simultaneously pumping out venom. This action can continue independently even when the sting and its sac of venom have been ripped from the body of the bee, which then dies.

Other organs inside the abdomen of a worker bee include a honey stomach where nectar is stored for the journey back to the nest, a true stomach where honey and pollen are digested, and a balloon-like rectum that can expand massively to hold waste – often for weeks at a time in winter – before it can be excreted. A queen's abdomen contains large ovaries that produce up to 2,000 eggs a day, and a drone's abdomen contains a store of sperm, along with elaborate sexual organs that are pushed out of the abdomen for mating.

The abdomen contains various other organs, including a long, stringy heart, all of which are suspended within watery haemolymph – the bee's equivalent of blood.

Hairs

Honey bees are extremely hairy, with up to three million fine hairs on their half-inch-long bodies. They even have hair on their eyes, sprouting from between each of the lenses. All of this hair is mainly for the collection of pollen, which provides the protein needed to raise the larvae. Incredibly, bees use electricity to help hoover up the pollen that they seek; a positive charge builds up on them as they fly, while the pollen in flowers is naturally negatively charged. When a bee delves into a flower, the pollen sticks all over their body, in much the same way as a balloon rubbed on a woolly jumper sticks to a wall. Efficiency is improved by the design of the hairs – they are plumose, meaning they are branched or feathery, giving the pollen lots of places to lodge. Worker honey bees are in effect flying feather dusters.

Legs

The front and middle legs of worker honey bees have rows of short, stiff bristles, called pollen brushes. In a feat of dexterity that would impress any yoga master, a bee covered in pollen will brush her legs over her entire body, combing pollen from hairs, wings and eyes. There is even a perfect-sized notch in the front legs for scraping pollen from the antennae. The collected, dust-like pollen has a little nectar added to it, sticking it together to make it less flyaway. It is then passed from one brush to another, continually gathering and sticking it together. The pollen is passed to the middle and then back legs where it is further combed and compressed before being fed into the pollen press – a clamp built into one of the hind leg joints. When the leg is flexed, the pollen is squeezed and pushed onto the upper part of the back leg. This area, called the pollen basket, is wide and flat with a fringe of stiff hairs and one very long hair in the middle. The pollen is moulded around the long hair, which acts as an anchor. Bit by bit, the gathered pollen is added until it forms a pellet enclosed by the hairs of the pollen basket. In this way, bees build an easily transported pellet of pollen on each of their back legs, weighing up to about one-third of their body weight. These pellets are flown back to the nest where the pollen is used to feed the young.

Honey bee feet have claws and a sticky pad that allows them to walk up objects and even upside-down across a ceiling. Their front feet have tiny sensilla hairs which detect sugar when they land on a flower, enabling bees to taste with their feet. A queen's feet have glands which secrete a pheromone that tells worker bees exactly where she has been walking in the colony, and how long ago, allowing them to locate her and even helping them to assess her condition.

ABOVE Not the claws of a prehistoric beast, but the foot of a worker honey bee.

The honey bee colony

Honey bees live in tight-knit family groups called colonies. Unlike many animals, including the solitary bees described later in this book, they are eusocial; they cannot survive as individuals outside their family group and are entirely dependent on one another to find food, raise their young and build a place to live. Honey bees are so interdependent that their colonies are classed as superorganisms – although they comprise tens of thousands of individuals, they operate like a single being.

The life of the honey bee superorganism revolves around a nest of combs that are built and maintained by members of the colony. Made from thousands of exquisitely constructed hexagonal beeswax cells, the combs are at the heart of everything the colony does; they are where eggs are laid, young are raised, and food is processed and stored. The ability to store food is key to the success of the honey bee colony, because it is this that enables it to survive when no fresh food is available, even through the harshest of winters.

Honey bee colonies are perennial, surviving from one year to the next and altering their size and characteristics in response to climatic and environmental conditions. In summer, colonies grow and flourish. The bees work to expand their nests, raise thousands of young, and collect and store pollen and honey for both immediate and future use. In winter, colonies shrink. The bees cease raising young, they live on the food stored in summer, and will stay in the hive and cluster together to keep warm. In spring, work begins again. The bees collect food to replace what was consumed over winter and to raise more young. When colonies have grown and conditions are suitable – usually in early summer – they reproduce, dividing themselves to create new colonies that will leave the nest and set up home elsewhere. This process is called swarming.

The dynamic life cycle of the colony, with its seasonal ebb-and-flow of raising young, producing honey and swarming, is part of what makes honey bees such a fascinating subject, and beekeeping such a captivating activity.

ABOVE Newly drawn comb with eggs in some of the cells.

Members of the colony

Honey bees spend their lives working for the survival and success of their colony. This is their only goal – there are no individual ambitions or egos. The day-to-day activity of each bee depends on its status, age, and position within the colony, as well as the specific needs of the colony at any particular time. A healthy colony at the height of summer might contain 50,000 or more individuals, comprised as follows.

The queen

Under normal circumstances, there is only one queen in a honey bee colony. She is usually the mother of the many thousands of bees that she shares the nest with. Her chief role is to lay eggs that will develop into new bees. Although nominally the head of a colony, the queen is little more than an egg-laying slave. Decisions about how and when she lays, and the number of eggs that she produces, are made not by the queen, but by the workers. What is more, when a queen is thought to be slightly less than perfect, or perhaps a little too old for the job, she can be ruthlessly dispatched and replaced.

Workers continually check the condition of the queen by monitoring the pheromones that she secretes. These pheromones, called queen substance by beekeepers, are complex chemical signals that inform workers about the state of their queen. If the pheromones are plentiful and balanced, life in the colony will be harmonious and productive. If the pheromones are insufficient or imbalanced, the colony can be bad-tempered, dysfunctional and likely to replace their queen.

A delicate balance of power exists in a honey bee colony. Although its ability to function and survive depends upon the queen, the workers – who are her daughters – won't hesitate to replace her should she fail to meet their performance expectations.

Workers

The vast majority of bees in a colony are workers. These are all daughters of the queen and, as their name suggests, they do all of the work. Within minutes of emerging from the cell in which they have developed, young workers busily begin doing everything necessary for the colony to survive and thrive. More than a dozen jobs are done by workers according to their

ABOVE A queen bee surrounded by her retinue of workers.

ABOVE A worker honey bee.

ABOVE A drone.

age, changing every few days as their maturing bodies become capable of other activities. For example, a newly emerged worker's first job is to tidy the nest, removing bits of wax or debris from the hive. A few days later she will start eating pollen and honey to nourish newly developed glands in her head. These glands produce the food that she will feed to the larvae. At about 10 days old, workers develop glands in their abdomen which secrete liquid beeswax. These workers then become responsible for producing wax and sculpting the hexagonal cells that are used for raising young and storing food. Other jobs include caring for the queen, capping and uncapping cells, processing nectar, removing dead bees, heating cells containing larvae, and processing and storing honey.

After about three weeks, the worker's sting hardens, becoming strong enough to be used as a weapon. At the same time, a gland in her abdomen begins producing venom. The worker now has both weapon and ammunition and is capable of stinging to defend the colony, and she begins to guard the entrance of the nest, flying around its periphery to detect possible threats. Gradually, she learns to navigate the surrounding landscape, memorising the way there and back across a radius of several kilometres. Now that she can venture far and wide, the worker begins seeking out, collecting and bringing home the resources the colony needs to survive.

In summer, worker bees spend about three weeks after emergence doing jobs inside the nest as described earlier, and are known as house bees. Once they have started flying, they spend the following few weeks collecting resources and are known as foragers. They live for a total of approximately six weeks before dying, usually of exhaustion. Workers produced in late autumn are physiologically different, and are able to live as long as six months in order to maintain the colony throughout winter and nurture the first generation of bees produced the following spring, after which they die off quite rapidly.

Drones

Drones are the males of the species. Their only job is to mate with virgin queens, inseminating them with their sperm and ensuring the survival of their genetic line. In mid-summer, when a colony contains perhaps 50,000 individuals, several hundred to a thousand might be drones. After emergence, young drones stay in the nest for about two weeks, gaining strength and becoming sexually mature. They don't work during this time, and are fed and preened by their worker sisters. Although they perform no physical tasks, the presence of plenty of drones seems to help to keep colonies stable and content. A lucky few drones succeed in fulfilling their life's purpose by mating with a queen; their reward is to die quite dramatically in the process. Those who do not mate survive only until the autumn, when their sisters, unwilling to feed and care for them throughout winter, expel them from the nest to die.

Reproduction in the honey bee colony

All honey bees start life as an egg. How and where that egg is laid, and the way in which the hatched larva is treated, will determine the sex and status of the resulting adult bee. It's a fascinating process that is at the heart of everything the colony does, as well as how the beekeeper manages and cares for it.

Honey bee queens have two ovaries within their abdomen. Each has about 200 ovarioles – tubes in which, during spring and summer, eggs are continually produced. Eggs leaving the ovarioles pass the spermatheca, a vessel in which millions of live sperm can be stored for several years. Most eggs passing the spermatheca will collect sperm before continuing to the ovipositor, the tube through which the queen lays her eggs. Lowering her abdomen to the bottom of an empty wax cell, the queen deposits a single, fertilised egg. She then moves on to the next empty cell to repeat the process. Incredibly, she might do this 2,000 times a day. It's somewhat awe-inspiring to think that each day in

BELOW Newly laid eggs.

early summer your hive might contain thousands more bees than it did the day before.

Not all eggs laid by the queen are fertilised. When the queen finds cells that are slightly larger than the others, usually towards the lower edges of the comb, she lays an egg but withholds the sperm – making the egg infertile. Incredibly, both fertilised and unfertilised eggs are viable. Fertilised eggs will develop into female bees, usually workers. Unfertilised eggs will develop into male bees – the drones (which, being larger, need bigger cells to grow in). This ability of a female to produce new life without the contribution of genetic material from a male is called parthenogenesis, and is possible in only a few animals. It means that whereas a worker bee has both a mother and a father, a drone has no father and will never have any sons. But the honey bee facts of life are yet more incredible because, although all fertilised eggs will hatch into female larvae, the vast majority of which will be workers, in certain circumstances a tiny minority will become queens.

All honey bee eggs take three days to hatch. The tiny C-shaped larvae lie on the base of their cells and are fed and tended by young worker bees in the nursing stage of their career. After eating honey and pollen, these workers produce secretions from two glands in their head. Mandibular glands produce royal jelly, a rich, white substance; whereas hypopharyngeal glands, in the top of their head, produce worker food, a less rich, clear substance. Another type of food is made from the contents of the worker bee's crop; this is a combination of nectar, honey and pollen, known as yellow food. The three types of food, collectively known as brood food, are mixed in varying proportions and fed to all larvae for the first two days after they have hatched. From the third day, worker and drone larvae are fed much less royal jelly and more worker food, with increasing amounts of yellow food. However, in certain circumstances a few female larvae will continue to be fed with very large quantities of mostly royal jelly. This very rich diet will cause these larvae to develop not into workers, but into queens.

Over about a week, each larva is visited and fed by workers several thousand times. The larvae fatten rapidly until they can no longer fit curled up in the bottom of their cells, and move to fill the cell lengthways. They then spin a silk cocoon around themselves, inside which they will pupate. At this

ABOVE Brood at different stages including larvae of various ages and capped cells. The cells at the bottom contain colourful pollen.

ABOVE Capped worker brood.

time, worker bees cap the cells with wax so that pupa development can continue undisturbed. Cells containing worker pupae are sealed on the ninth day after their egg was laid, and drones on the tenth.

Inside their cocoons the pupae undergo metamorphosis, shedding their skins several times until they are transformed into fully developed bees. They then nibble their way through the wax capping of their cell and emerge into the hurly-burly of the nest. Worker bees emerge 21 days after their egg was laid, and drones after 24 days. When the cells have been vacated, workers clean them out and prepare them for the queen to lay another egg.

The development of queens

As mentioned earlier, the development of new queens is slightly different to that of workers and drones. It is important that beekeepers understand the differences because they have an effect on how honey bee colonies are managed.

There are several reasons why worker bees might want to produce a new queen. Primarily, it is because they intend to swarm. When they do swarm, the old queen will leave the nest along with thousands of bees. They usually leave behind one or more replacement queens who are still in the process of development. (See pages 80–91 for more on swarming.)

ABOVE Capped drone brood.

ABOVE Emerging worker.

HONEY BEE DEVELOPMENT

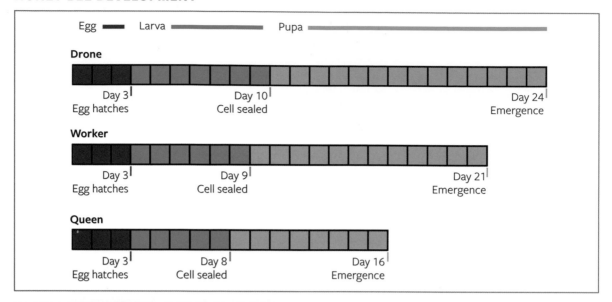

Egg ▬ Larva ▬▬▬ Pupa ▬▬▬▬▬▬▬

Drone

Day 3	Day 10	Day 24
Egg hatches	Cell sealed	Emergence

Worker

Day 3	Day 9	Day 21
Egg hatches	Cell sealed	Emergence

Queen

Day 3	Day 8	Day 16
Egg hatches	Cell sealed	Emergence

ABOVE An emerging queen.

New queens can be produced in the same, small hexagonal cells that are used to raise workers. Indeed, any fertilised egg laid in an ordinary cell has the potential to become a queen. But if a colony wants to swarm, workers will first prepare special cells called queen cells, or swarm cells. These are attached to the surface of the comb, and are not oriented horizontally like the cells of the comb, but vertically, with their entrance at the bottom. When first built they look a bit like acorn cups, and are known as queen cups or play cups. When extended to accommodate growing queen larva, they look a bit like peanuts in the shell.

When the queen finds a prepared queen cup, she will lay a fertilised egg inside it. The female larva that emerges will be fed huge quantities of brood food that is extra rich in royal jelly, to make it develop into a queen. Queen larvae grow very rapidly, the workers extending the wax walls of the cell to accommodate them as they do so. On the eighth day after its egg was laid, the queen larva begins to pupate. The workers seal the cell and, because they know a viable new queen is now sealed in her cell and going through metamorphosis, they see this as their opportunity to swarm. Perhaps half the workers will leave the colony, taking the old queen with them. New queens emerge from their cells 16 days after their egg was laid (usually about eight days after a swarm has departed) and, depending on the exact circumstances, may then become the new head of the colony. However, as with most things honey bees do, there are alternative scenarios.

A prime objective of most beekeepers is to prevent their bees from swarming. Therefore, it is important to be able to recognise the signs of swarm preparation and understand the timings of brood development, particularly that of new queens (see table above).

Queen mating

Newly emerged queens are infertile because they haven't yet mated. For the first few days of her life, the virgin queen looks after herself and is largely ignored by her sister workers. At about a week old, the virgin queen will fly from the colony to mate. Upon her return, the mated queen becomes the centre of the workers' world, and everything they do pivots around her.

Honey bee mating is a spectacular event which few beekeepers ever witness. It is a process that is only partially understood and there is much yet to discover. Mating usually takes place on warm days in early- to mid-summer. Drones from many colonies leave their nests and fly perhaps 5 miles (8km) to a drone congregation area (DCA), although there is evidence that mating can also happen very close to the apiary. At DCAs, drones gather in large numbers, flying high in the air, possibly sometimes resting on vegetation, waiting for virgin queens to arrive. When a queen does arrive, the drones detect her pheromones and give chase. The fastest drone will catch up with the queen and mount her. His endophallus, an elaborate arrangement of sexual organs, enters the queen and is then squeezed so tightly by her abdominal muscles that the drone is thrown backwards, his reproductive organs are ripped from him, and he falls to the ground and dies. Parts of his endophallus remain in the queen and continue to transfer sperm. The process is repeated several times, each subsequent drone first having to remove the remains of the previous suitor before he himself can mate. A queen mates with perhaps 20 drones, a proportion of their sperm being mixed and stored in the queen's spermatheca, where it is kept alive until needed. Mating with so many drones, probably from multiple colonies, ensures genetic diversity, which is thought to help promote colony resilience and adaptability. Mated queens return to the hive and become productive egg layers, usually within a few days after mating.

ABOVE Love is in the air; a rarely seen drone congregation area (DCA).

Becoming a beekeeper

What kind of beekeeper will you be?

If you join a local beekeeping association, or social media chat group, you will soon discover that although many people love honey bees, few can agree about the best way to look after them. Perhaps more than with any other kind of livestock or pet, the way that people approach keeping bees can be highly individual, with a great many options available. The kind of beekeeper that you become will be influenced by the books you read and the beekeepers you meet. It will also reflect your personal beliefs, ethics and aspirations, as well as your living arrangements, available spare time and attitude to expenditure.

Until the late nineteenth century, most honey bees were kept in skeps – straw baskets in which swarms captured in spring were housed and maintained throughout the summer. As winter approached, the bees were often killed to allow the harvesting of their honey and wax. These techniques remained fairly common until the 1930s. In the second half of the nineteenth century, the movable-frame hive was invented. This enabled colonies to be kept in standardised wooden boxes, and their wax combs to be held in wooden frames, easily removable for the harvesting of honey. It also allowed colonies to be managed to prevent swarming, and to be closely inspected and treated for disease. The method was based on the natural needs and habits of honey bees, but was designed to make it easy for humans to manage them and harvest their produce. Despite regional differences, slight variations in hive design and countless individual methodologies, the moveable-frame method became dominant throughout much of the world. Broadly, it is the way that most people keep their bees today and is what is usually thought of as conventional beekeeping.

Recently there has been increasing interest in other methods of beekeeping. Natural, alternative or bee-centred beekeeping methods try to put the needs of the bees at the heart of everything that is done. There is a wide range of approaches that might fit these descriptions, and no two natural beekeepers (or conventional beekeepers, for that matter) do things the same way or for the same reasons. Most people who are interested in these ideas use hive designs that they believe allow bees to live and behave in a more natural way. Some hives are simply hollowed logs, while others are made of straw or other materials and allow the bees to build their comb in whatever configuration they want. Some, such as the horizontal top-bar hive or the Warré hive, impose a little more order, giving the bees wooden bars as a starting point for the construction of their combs. Methods like these are usually designed for minimal disturbance or management by the beekeeper. Most natural beekeepers see productivity as a low priority, so their methods tend to allow harvesting of only small amounts of honey – if any. Such methods don't always allow colonies to be easily inspected for pests or disease. Indeed, many natural beekeepers believe that bees should not be treated at all for disease, but be allowed to live or die in a 'survival of the fittest' approach known as Darwinian beekeeping. Similarly, feeding bees sugar is often frowned upon, with colonies that are unable to survive solely on what they have naturally gathered often perishing. Natural beekeeping methods often allow for the unmanaged swarming of colonies. This might be natural and possibly beneficial for the bees, but is unlikely to improve relations with the neighbours.

Most natural beekeepers are extremely knowledgeable, thoughtful and caring. They are often very experienced at keeping bees using conventional methods, and have gradually developed alternative approaches that they

OPPOSITE PAGE Removing the summer honey harvest.

ABOVE Skeps were made to different designs according to local tradition.

feel work better for them and their bees. However, natural beekeeping methods can also appeal to new beekeepers who want to keep bees as part of a certain kind of lifestyle, but who often have little understanding of the basic requirements of honey bees, or of the responsibility that comes with the privilege of keeping them. Occasionally, natural beekeeping can be an excuse for laziness or negligence.

In this book, I concentrate on what are generally thought of as conventional beekeeping methods because I believe that these are the best way for new beekeepers to learn about the biology, behaviour and needs of honey bees. Much of the pleasure of beekeeping comes from the close observation of bees and the ever-changing condition of their colonies. This is something that conventional beekeeping allows in a way that alternative methods often do not. Using conventional hives, equipment and techniques enables a beekeeper to study their bees closely, to see when a queen is laying eggs, or if a colony is preparing to swarm. It is easy to see when bees are accumulating stores of honey and pollen, and relate that to the weather or what plants are currently flowering. The beekeeper can look for pests and diseases, learn to recognise and treat for them if necessary, and feed the bees if they notice that there is a danger of starvation. Conventional methods allow the beekeeper to manage colonies in various ways and to

see how those changes help, or possibly hinder, a colony. Mistakes can be spotted and rectified, suggesting how things might be done differently in the future. Often it is a case of learning when it is best to leave the bees alone, allowing them to thrive without the very close attention you might lavish on them as a beginner.

Although natural beekeeping methods are claimed to be in the best interests of the bees, conventional beekeeping also strives to keep bees as healthy and contented as possible, with ethically based decision making – for example, much honey to harvest, how to approach the treatment of disease and whether to buy imported or locally produced queens. These will all be influenced by your feelings, politics and beliefs.

However you start keeping bees, and whatever methods and equipment you first adopt, it is certain that in five years' time you are likely to be a very different kind of beekeeper. By then you will have let a few swarms get away, lost a few colonies to disease or starvation, bought and discarded plenty of pointless gadgets, and found out how other beekeepers do things differently. You will gravitate towards methods that best suit you and your bees. There are some rights, some wrongs, and plenty in between that beekeepers will always disagree about. Beekeeping covers a very wide spectrum, and everyone takes a little time to work out exactly where they fit on it.

Bees in your garden

Most beekeepers keep their bees in their garden. This is both practical and enjoyable. It means that you don't have to travel far to tend your bees, you can easily fit beekeeping around your other domestic activities, equipment is always to hand, and you can pop out to see the bees whenever you feel like it – which is likely to be often. If you place a comfy seat near your hives, you can sit and watch the bees at work. It's a glorious way to spend a sunny hour or so.

Some people find that their garden isn't suitable for keeping bees, or that their hobby outgrows the available space. In such cases, apiaries might be established elsewhere. Many beekeepers also have what are called out-apiaries because, for various management reasons, it is useful to keep bees in more than one location. Other good places to keep bees include other people's gardens, nearby farmland, allotments, parks, churchyards, historic properties, or even on the flat roofs of commercial buildings.

Neighbours

You might find the idea of keeping bees in your garden exciting, fascinating and charming, but your neighbours may not share your enthusiasm. If you have a large garden and distant neighbours, you need not be too concerned about the potential impact of your bees – it's likely no one will even know they are there. If you have a smaller garden with nearby neighbours, particularly if you share a boundary, there is more to consider.

In the UK, there are no laws that prohibit or restrict beekeeping in gardens (in other countries there can be), but you should consider carefully whether your new hobby might cause distress or possibly even harm to your neighbours. Simply knowing there are hives next door could cause some people extreme anxiety.

Most people assume, incorrectly, that if their neighbour keeps bees they will get lots of bees in their own garden. In fact, most honey bees when leaving the hive will quickly fly upwards and straight towards their pre-planned foraging destination, which might be some miles away. On their return, they will go directly back

ABOVE Comb being removed from a horizontal top-bar hive.

ABOVE A beehive in a central London garden.

colony might become excessively defensive and begin stinging for no apparent reason. In such cases the beekeeper needs to be able to act promptly, usually by either moving the bees to another location or replacing the queen (see page 99).

Swarms, though a fascinating, benign and fleeting occurrence, can be a nuisance to your neighbours. Swarming bees are scary to those who don't understand them, and it can be worrying or inconvenient if a swarm moves into a chimney or a loft. Many neighbours will have no objection at all to the bees living next door – or even know that they are there – until they decide to swarm. Some strains of bee are less 'swarmy' than others, but beekeepers, particularly in urban areas, must manage their bees' natural instinct to swarm (see page 97).

If you have nearby neighbours, particularly if they overlook your garden and will be able to see your hives, it is courteous to speak to them before you set up your apiary. Many people are fine with the idea of having bees next door, and even those who are initially unsure will often come around to the idea once they understand more about what is involved, are kept informed and are seduced by the prospect of the occasional jar of very locally produced honey. When you have a little more experience, you might even invite your neighbours to come and see inside a hive and learn more about the fascinating creatures living next door.

How many bees?

Many people who are thinking about having bees imagine that they will have a single hive – just enough to provide some interest, pollinate garden plants and produce a few pots of honey. In fact, it is difficult to maintain a single colony of bees because there are times when it is necessary to borrow resources from a healthy colony in order to support an ailing one. It's fine to start with one colony, but you should aim to have at least two in your second year of beekeeping. But be warned, many beekeepers who start with modest ambitions soon find they have far more colonies than they ever intended – beekeeping is highly addictive. It's a good idea to set your limits and stick to them.

into the hive. If your neighbour already has flowering plants in their garden, they probably already receive visits from local bees. Any bees visiting from next door will hardly add to that number. It is true that your neighbours might have more bees flying *over* their garden, but this can be managed to avoid disturbance (see page 43). It is possible that your bees might visit your neighbour's garden in quite large numbers if they find a suitable source of water there. Again, this can be addressed before it becomes a problem.

There is a chance that your neighbours might suffer the occasional sting, and this is unlikely to be well received. The risk can be minimised by keeping gentle strains of bees, by the thoughtful placement of your apiary, and by choosing appropriate times to open hives and inspect your bees. Occasionally, an individual

RIGHT Beehives in a modest-sized, well-planted urban garden.

How to get started

Establishing an apiary

When deciding where to keep bees, you need to consider where will best suit you, where will best suit your neighbours and, crucially, where will best suit the bees. It's rare to find the perfect spot, and compromises will usually be necessary. If you can't find a good place in your garden, or your family aren't agreeable, you may have to look for somewhere else to keep your bees. Choosing the right location can make a big difference to the success of your new venture, and it can be difficult to make changes if you get it wrong.

First, choose a location that will allow family life to continue as normal. I have seen hives placed outside patio doors so that the bees could be viewed from the comfort of the living room. This must have been more absorbing than watching TV, but it was far from ideal. Select a spot in the garden that is some distance from the house, ideally a little tucked away and not too near frequently used areas like patios, paths or washing lines. The last thing you want is family members being afraid to use the garden in case of unwanted encounters with bees.

As the beekeeper, it is important to find a place that works well for you. You will spend quite a bit of time around your hives and will need to access them easily and have enough room to work. When inspecting colonies, you will require room to stand next to the hive and room on the ground to place various pieces of equipment. A minimum space of about one metre surrounding each hive is ideal.

Consider whether your proposed spot might cause disturbance to your neighbours. As a rule, it is better if they cannot see your hives – out of sight being to some degree out of mind. A bigger issue is that bees travelling to and from their hives are likely to fly over your neighbour's garden. This could cause some consternation, even though the worst outcome is likely to be a fast-flying bee bumping into someone's head.

ABOVE An obstruction in front of a hive can force bees to fly high over your neighbour's garden.

OPPOSITE PAGE An apiary in a rural location with trees and hedgerows offering abundant forage.

Locating hives at the far end of your garden can be a good idea as it means that bees are likely to fly mostly over the bottom of your neighbour's garden – usually the least visited part.

Bees leaving hives typically ascend gradually, levelling off at a height of several metres, well above head height. If you can force them to fly up to cruising altitude almost as soon as they leave the hive they will fly over your own garden, and that of your neighbour, without getting in anyone's way. On their return, they will fly at high altitude until they drop down to their hive in the last few metres. Sheds, wooden or mesh fence panels or shrubs placed a few metres in front and to the sides of a hive will force bees to leave and approach at a steep angle. Somewhat counter-intuitively, if your neighbours are bothered by the idea of your bees, placing the hives near a shared fence or hedge can be the best solution because it will force the bees to fly at a height that keeps them out of everyone's hair.

ABOVE An apiary in a woodland clearing.

A garden shed can be repurposed as a bee house, the hives being kept against the inner walls and the bees flying to and from entrances cut in the building's sides. Bee sheds offer several advantages. They help to disguise the fact that there are bees in your garden, making neighbours less anxious and making theft or vandalism less likely. Hives in sheds can be opened for inspection on cooler, windier days than hives outside – days when neighbours are less likely to be using their garden. When you open hives, bees will fly up into the shed and towards a window, which should be left open to allow their escape. They will quickly re-enter the hive using the front entrance. This means fewer potentially angry bees flying around the garden and possibly bothering neighbours. Keeping hives in a shed also allows spare equipment and tools to be stored nearby.

Keeping bees on allotments

If you keep an allotment, you might be able to house your colonies there. Until recently, livestock of any kind were rarely allowed on allotments, but rules tend to be a bit more relaxed now – though you will probably still need to make a case to the allotment committee or local council. Reasonably, some councils ask that anyone wanting to keep bees on an allotment should have several years' experience, proof of attending a course, or will have attained the BBKA Basic Certificate – an entry-level certificate that can be obtained by taking an examination after keeping bees for one full year.

The main concern, of course, is stings. This is a possibility but, as when keeping bees in small gardens, precautions can be taken. If your plot is on the edge of the allotments, bees can easily be directed away from the rest of the plots by facing hives in the opposite direction. Otherwise, you might be able to secure a small area at the edge of the site for your hives. Surround the apiary with a high hedge or fence – plastic windbreak material works well – if only until something else can grow. This will force the bees to fly upwards when leaving the hive, avoiding passing gardeners.

The key to success is educating fellow allotmenteers and keeping them onside. Arrange a meeting to address concerns and talk about how beekeeping works. The important points to explain are: what happens when bees swarm; the fact that bees may gather around sources of water; when and why bees might sting (and that they are highly unlikely to do so when visiting flowers on the allotment); and what actions should be taken in the event of being stung. You should also point out that the quantity and quality of produce grown on allotments has been shown to improve significantly when honey bees are kept nearby. Finally, remind everyone that there will be free (or perhaps discounted) honey on offer – often a remarkably effective inducement.

Make sure the bees you keep are gentle, removing colonies or re-queening them if they show signs of being defensive. Try to do your inspections in good weather, and on weekdays, when fewer people are likely to be working on the allotment.

ABOVE Hives on an allotment.

THE BEST PLACES FOR BEES

Free-living honey bees put a lot of effort into finding the right spot to live, as it can make all the difference in their struggle for survival. Keeping bees in hives, often in close proximity to one another, is an unnatural situation, so when choosing an apiary site, it is important to try to provide as many of the bees' preferred living conditions as possible.

Wind: One of the most important considerations is shelter from the wind. Bees have difficulty taking off, flying and landing in strong winds, so having hives sheltered by nearby buildings or vegetation helps. If you can face hives away from prevailing winds, all the better. Colonies in hives sheltered from wind will spend less energy maintaining their temperature in winter.

Sun: Colonies can benefit from the warmth of the sun in winter and spring, but full sun in the summer can make a lot of work for the bees, who might need to ventilate the hive to keep their brood nests within the ideal temperature range of 32–36°C. Some say that overheated colonies are more likely to swarm. The best solution is to site hives under deciduous trees so that they benefit from sun in winter but are partially shaded in summer.

Orientation: If they can, wild honey bees will occupy places with south-facing entrances. This allows them to benefit from long hours of sunshine and means a longer working day. South facing is best for hives in your garden, although if the other criteria are met this is less important.

Space: Wild colonies tend to keep their distance from one another, which is quite different from the way we force them to be close neighbours. One problem with hives being very close together is that bees can sometimes use the wrong front door and enter another hive – known as drifting. Drifting can spread disease and encourage colonies to rob honey from one another. Well-spaced hives placed at differing angles rather than in straight lines can help to avoid this.

Be aware that allotments are easily accessible places and that theft and vandalism are possible. A secure, high fence and lockable gate are therefore a good idea, or alternatively you could surround your hives with a spiny, flowering hedge that is also useful to the bees, such as Darwin's barberry (*Berberis darwinii*).

Choosing a beehive

There are many types of beehives to choose from and it can be quite bewildering when you first begin looking. In fact, conventional hives are all more or less the same, with only minor differences in size and some features.

The first widely used conventional hive was designed by American preacher Lorenzo Langstroth (1810–1895), though he relied heavily on the work of earlier pioneers. The idea was that bees would build their combs in removable wooden frames which were hung inside a wooden, box-like hive. This would enable combs to be easily removed so that honey could be extracted, or to be swapped between hives for various management purposes. However, there was a snag with this concept. If there was a narrow gap between the edge of the wooden frame and the internal walls of the hive, the bees would fill it with sticky propolis (see page 108). This stuck the frames into the hive and made them difficult to remove. If a wider gap was left between the frame and walls, the bees instead filled it with wax comb (known as brace comb), which also fixed the frames in place. It had already been noticed that when wild honey bees build their nests (inside a hollow tree, for example) there is a more or less consistent-sized gap between each comb. Also, wherever combs are attached to walls, bees leave similarly consistent-sized passages to walk through. These gaps are 6–8mm. Langstroth found that if gaps of this size were incorporated around frames as well as any other spaces within the hive, the bees wouldn't fill them with propolis or brace comb, and frames and other parts of the hive could therefore easily be moved or rearranged. The 6–8mm gap became known as 'bee space, and it made modern beekeeping as we know it possible. Langstroth incorporated bee space into his hive design, which he patented in 1852, and the Langstroth hive is today the most widely used hive in the world.

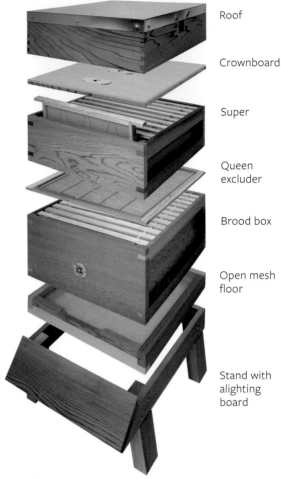

Roof

Crownboard

Super

Queen excluder

Brood box

Open mesh floor

Stand with alighting board

ABOVE A National hive and its component parts.

ABOVE An elaborated Langstroth hive from 1859.

ABOVE WBC hive.

ABOVE Polystyrene National hives.

In the UK, various hives evolved that made use of Langstroth's ideas. The best known is a hive designed around 1890 by William Broughton Carr (1836–1909). His hive was similar to Langstroth's, but used an additional outer layer of boxes to surround the inner boxes in which the bees lived – what has become known as a double-walled hive. The gently downward sloping outer boxes of the later version of Carr's hive are called lifts, and they stack on top of one another to produce the pleasing stepped pyramid appearance that most people consider to be the classic beehive shape. Carr named the hive after his own initials: the WBC hive.

The WBC is the quintessential English beehive, but it is less popular today because, although perfectly adequate, it is expensive to buy and somewhat cumbersome to use. Most British beekeepers now use a single-walled hive called the National. This originated at about the time of the First World War, when there was a need for hives that were cheaper and easier to make. The current version of the design gained popularity after the Second World War and is properly called the Modified National, though most people call it simply the National.

Other hive models are available, with some being more common regionally than others, but they are all more or less of the same design. The main differences between each type are the dimensions of the frames and the internal volume of the various boxes in which the bees live and work.

When choosing a hive, you might select the WBC if aesthetics are important, or either the National or the Langstroth for utility. Langstroths are of simpler design and construction, so tend to be the cheapest. WBCs and Nationals use the same-sized frames and, since the majority of UK beekeepers use these sizes, either of these hives is a practical option. Using common-sized frames means that they are readily available and cheap, but also that you are more likely to be able to buy colonies of bees on frames that will fit into your hive. Importantly, should you need a frame of brood and eggs in an emergency (which is not uncommon for a beginner) you are likely to know local beekeepers with the same hive type who will be able to help. If you join your local beekeeping association, it's a good idea to ask members what hive types are common in your area and what they would recommend.

Modern hives

Recent innovations in hive design have resulted in some models that are less traditional in appearance, and are made of modern materials. The usual claim is that these hives make beekeeping simpler, easier in small spaces and less time-consuming – but rarely cheaper. The Flow hive is the best known of these innovations. The design is very clever, and in the right conditions (such as in its native Australia) it can work well. In the UK, however, some of our honeys can be too thick to work well with the system, even setting solid and glueing up the mechanisms. I recommend starting with traditional hive types, and experimenting if you wish to do so when you become more experienced.

Hive anatomy

Whether you choose the WBC, National, Langstroth or any of the other available options, all conventional hives consist of the same basic parts. Essentially, beehives are just a stack of boxes and other quite simple components, put together in a particular order. You will add or take away sections as part of the beekeeping management routine. Note that in the following section, various hive measurements are quoted in inches, with metric measurements given in parentheses. This is because beekeepers and suppliers commonly use inches when referring to the size of certain equipment, particularly frames.

Floor

Hive floors are simple tray-like structures with an opening on one side that forms an entrance for the bees. There are various entrance designs, but the most common one is simply a gap that is about the width of the hive, which the beekeeper plugs with a removable wooden entrance block. The entrance block, a long strip of wood with variously sized holes in it, can be turned around to alter the width of the bees' front door. Floors traditionally had a solid wooden base, but now usually incorporate a sheet of wire mesh with a removable tray underneath. The mesh allows varroa mites (see page 143) to fall from the hive onto the tray, which can be removed so that their numbers can be monitored. In most hives the floor is separate to the hive stand, but in the case of the WBC, the floor and stand are one piece.

Brood box

When bees enter a hive, they move up into the brood box. This sits above the floor and is usually the largest of the boxes used in a hive. It is where the queen lives and lays her eggs. In summer, the frames and combs in this box are used mostly for brood rearing. In winter, combs in the brood box are filled mainly with stored honey for consumption during the cold months.

The size of the brood box is one of the main variables between different hive types. If you live somewhere with a mild climate and have prolific bees (for example, Italian-type bees – see page 59), a large brood box is a good idea. Conversely, if you live somewhere cooler or have less prolific bees (such as British native types), a smaller brood box is preferable. Therefore, it is a good idea to decide how you will acquire your bees, and what kind they are likely to be, before you buy your hive equipment.

One reason National hives are commonly used is that they offer two sizes of brood box. Originally, all National brood boxes had frames that were 14 inches wide and 8.5 inches deep (35.56cm × 21.59cm). However, with a warming climate and generally more prolific bees (because of imported strains), many beekeepers found they often had to add a second, smaller box (a super – see below) to give additional space for the brood nest. This system, still used by many beekeepers, is called brood and a half. In some cases, two full-size brood boxes will be used. The modern alternative is a larger brood box that holds frames 14 inches wide by 12 inches deep (35.56cm × 30.48 cm), giving about a third more room for the brood nest. So, with National hives you have a choice of two brood box sizes; the shallower but confusingly named British National (or British Standard) Deep, and the deeper National 14 × 12 (referred to as '14 by 12s'). Although the boxes inside a WBC are smaller because they are made of thinner wood, they use the same size frames as National hives, and you have the same choice of two brood box depths. Be aware that 14 × 12 frames when full of brood and stores are considerably heavier and harder to handle. Again, it is worth asking local beekeepers what they recommend, and handling different-sized frames to see which you prefer.

Queen excluder

Because queens are larger than workers, a barrier with bars or holes of the right width can be used to exclude them from certain parts of the hive. A queen excluder is placed on top of a brood box when additional boxes and frames are added over the top as a place for bees to store honey, preventing the queen from climbing up and laying eggs in them. Don't skimp on queen excluders – buy a framed wire queen excluder which consists of a wooden frame holding a grid of stainless-steel wire bars. They cost a little more than flimsy plastic or metal alternatives, but last longer and are better for both beekeeper and bees.

 Worker bee squeezing through a queen excluder.

ABOVE A 14 × 12 brood frame hanging from hive stand.

Supers

Supers are shallow boxes placed above a queen excluder, to give the bees additional space to store nectar and honey. They are usually added in late spring when plants begin to flower, and are removed in late summer when the honey is harvested. In a good honey year, you might add a second or possibly even a third super. Most hives when bought as a kit come with two supers, but buy one or two extras – it's worth being optimistic!

Crownboard

This is a simple wooden cover that goes over the topmost box in the hive – typically over a super in the summer or the brood box in winter. They have one or two central holes in them which can be used when feeding bees (see page 90), or when harvesting honey (see page 119).

Roof

This is the topmost part of the hive. It keeps the lower components dry as well as helping to hold them all down – hive parts are not fixed together in any way, and rely only on gravity and stickiness to keep them in place. Roofs can be flat-topped or pitched. Pitched roofs look more attractive, particularly in a garden. However, flat roofs are more practical – you can rest things on top of them, and when you remove them during inspections you can lie them flat on the ground and pile other hive components on top of them. Suppliers usually offer a choice of deep or shallow roofs, referring to the depth of the timber surround. Deeper roofs are heavier to lift from the hive but are less likely to blow off in high winds.

Stands

Hives need to sit on a stand of some sort. This keeps them away from damp ground and lifts them to a height that won't give you backache when you are working with your bees. As a rule of thumb, the top of the brood box should be at about waist height. Wooden stands of various designs can be purchased or are relatively easy to make – as long as they are strong enough to take the weight of a hive full of honey in the summer. Strong, treated timber can be used to make a simple framework, or a stack of breeze blocks will do the job. Stands can take a single hive, or several in a row. A popular design has room for two hives with a gap in between where roofs and other hive parts can be placed during inspections.

The legs of stands can sink into the ground and be prone to rot. It is best to place each foot on a tile or brick to spread the weight and keep them dry. Alternatively, lay a concrete paving slab, using a spirit level to make sure it is perfectly flat, and place the hive stand on top for a really firm, dry footing.

Construction materials

The best material for hives is western red cedar wood, which is light, durable and needs virtually no maintenance. One of the most delectable of beekeeping experiences is to lift the lid of a hive and savour the incense-like aroma of honey, beeswax and cedar. Some suppliers offer cheaper alternatives, usually made of pine. These need to be sealed, or painted and recoated every few years. They are heavier than cedar and will not last as long. Beekeeping certainly isn't cheap when you are setting up, but it really is worth buying the best that you can afford. A good-quality cedar hive will last a lifetime.

Polystyrene is now a popular alternative to wood. Poly hives have some advantages; they are lightweight and easy to lift and transport, they help to keep the bees warm in winter, and they are slightly cheaper to buy. Bees live very happily in poly hives, but whether the hives are attractive to the human eye is a matter of taste – some are certainly nicer looking than others. Poly hives need to be painted with masonry paint to prevent the material from degrading in sunlight. Nevertheless, they do become brittle with age, chunks can get knocked out of them, and they will not last nearly as long as wooden hives. Poly hives are made of non-renewable petrochemicals and will end up as landfill decades before cedar hives might need to be repaired or retired. In the meantime, trees will be growing, ready to produce the hives of the future. It's a personal decision, but if you are interested in keeping bees for ecological reasons, wooden hives would seem the better choice.

BELOW Cedar hives last for decades without any treatment, but a coat of raw linseed oil every few years will help keep them in good condition.

Building hives

To get started you need only one hive, but you will certainly need a second (or third) before very long. They can be purchased ready-made, but this is expensive. In the case of the finest WBCs with perfect dovetail jointing, this is the only option. However, most hives can be bought as flatpack kits for home assembly. This is a fairly simple job that can be done by anyone with basic DIY tools and skills. There are a few things that can go wrong, so study the instructions and watch the videos that are on the websites of many of the suppliers. The pieces usually slot together easily, but double check that everything is in the correct place before you start nailing and gluing, otherwise it can be very difficult to correct a mistake.

Frames, runners and bee space

Inside each brood box or super hangs a gallery of wooden frames. As their name suggests, they look rather like picture frames, with a handle (called a lug) at each end to lift and hold them by. Within the frames, the bees make their own beautiful artwork, sculpting the thousands of hexagonal wax cells in which they raise their young and store food.

You will need to buy the correct frames for your type of hive, and in two different sizes – deep ones for the brood box and shallow ones for the supers. Frames can be purchased ready-made, although you pay a premium to cover the cost of assembly and delivery. They are cheaper and easier to store if bought in kit form and assembled as needed. Different hive types take differing numbers of frames. A National hive needs 11 frames for the brood box and 11 for each super. It's worth buying more than you initially think you need as occasionally they break or have to be replaced, and all will need replacing or reconditioning every few years.

You will also need a dummy board – a thin, solid frame that takes the place of one frame of comb in the brood box. If you have a small number of frames in the box, the dummy board provides an internal wall, in effect making the hive smaller. If the hive has a full set of frames, a dummy board takes the place of one outer frame. This means that during inspections you can first remove the dummy board, making some space to easily move the brood frames around.

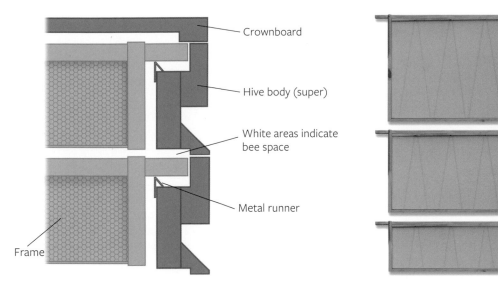

Crownboard

Hive body (super)

White areas indicate bee space

Metal runner

Frame

ABOVE Hive components fit together to provide interior passages known as bee space. This hive has bottom bee space.

ABOVE Three sizes of National frame: 14 × 12 brood frame; National deep brood frame and super frame.

ABOVE Assembling a frame.

ABOVE Sliding foundation into a frame.

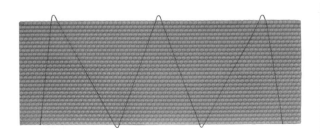

ABOVE Wired foundation.

RIGHT Tapered shoulders keep Hoffman frames perfectly spaced. They sit on metal runners.

ABOVE A National super. Rather than frame runners it is fitted with castellations. Frames slot into the gaps to maintain the correct spacing.

Frames in kit form consist of five pieces: a top bar, two grooved side bars and two bottom bars. The top bar comes with a built-in component called a wedge which needs to be removed before assembly and is later used to help secure wax foundation into the frame. To assemble frames, you need a small hammer, some nails of the right size, and pliers to remove the nails when something goes wrong. There are videos online that show you how to put the frames together, and the suppliers' websites usually have instructions too. No special DIY skills are needed, and once you get the hang of it they take only a minute or two each. The most important thing is to make sure that they are square. A set square is useful for checking that the corners are at right angles.

It is important that frames in a hive are accurately spaced. Too close to one another and the bees will stick them together with propolis, but too far apart and the bees will build brace comb between them. In either case, they will be difficult to remove from the hive. The best way to ensure correct spacing is to use self-spacing frames, known as Hoffman frames. These are designed so that when they are pushed against each other there is a very small point of contact and the gap between them is correct.

When placed in a hive, the lugs of the frames rest on narrow metal or plastic bars called runners. These allow the frames to be moved back and forth within the box, a bit like folders in a filing cabinet – although often they are too sticky with propolis to slide. Runners also hold the lugs above the surface of the rebated shelf of the box in which they sit, giving a bee space beneath them. If they sat directly on the rebate, the bees would weld the lugs to the hive with propolis.

In the case of National hives, runners are usually positioned so that when frames are resting on them, the upper surface of the top bars is flush with the top of the brood box or super. If you were to place a ruler over the top of the box it should touch all components and lie flat. When frames are positioned like this there will be a gap between their bottom bars and the bottom of the brood box or super. This gap will be one bee space wide. This means that when you stack boxes – putting a super on top of a brood box, for example – there will always be one bee space between the top of the frames in one box, and the bottom of the frames in the box above. This way the bees will not use propolis or brace comb to stick the frames above to the frames below.

If, as described, frames are level with the top of a box and have a space at the bottom, the hive is known as a bottom bee-space hive. However, sometimes it is the other way around and there is a space at the top rather than at the bottom, in which case it is a top bee-space hive. Experienced beekeepers have their preferences, but whether you have top or bottom bee-space hives does not really matter, other than that you must make sure all components you own are the same, either one or the other. Generally, National hives are bottom bee-space and Langstroths are top bee-space – but not always!

Foundation

Frames are usually fitted with a sheet of wax foundation before being placed in the hive. As its name suggests, this provides a foundation of wax on which the bees can build their comb. Bees can build comb without assistance from us, but foundation gives them a helping hand by supplying some ready-made beeswax footings, which they nibble away and recycle to help build cell walls. Foundation also ensures that the bees build their comb where we want them to.

Wax honeycomb is remarkably strong but is not designed by the bees to withstand being lifted out of a hive for inspection by beekeepers, nor to be placed into extractors and spun at speed to remove the honey. Most foundation therefore has embedded within it a thin strand of wire to provide additional strength. If you want to produce honeycomb that is eaten in chunks you can instead use very thin, unwired foundation in your super frames, or none at all (see page 123).

Foundation is sold in sheets that are the right size for each kind of frame. It is fitted into a frame by

ABOVE Three types of hive, each needing different sizes and types of frame. It is best to choose one and stick with it.

sliding it between the two bottom bars and along the grooves in the side bars. The three wire loops along the leading edge of the foundation are bent over and the sheet is pushed all the way to the top bar of the frame. The wooden wedge, removed from the top bar of the frame during assembly, is pushed up against the foundation and over the wire loops. Three small nails are then hammered through the wedge and through the wire loops to secure everything in place. Before use, foundation should be stored somewhere cool and dark and kept flat so that it does not buckle. Once frames have been fitted with foundation, they should be kept upright to avoid the wax warping. Always buy your foundation from reputable suppliers. Bargain foundation can be bought online, but it can contain other types of wax or be contaminated in other ways. Foundation should always be 100 per cent beeswax. Plastic foundation is available, but it is not a natural product and results tend to be variable.

Warm way or cold way

With square hives like the National or the WBC, the beekeeper must decide whether the frames will hang parallel to the front of the hive, or at right angles. These are known as the warm way and the cold way, respectively. As with so much else in beekeeping, the experienced will argue about which is best for the bees. I have used both ways successfully but generally use the cold way – meaning that if you were to look at the front of a hive you would see the frames end on, rather like the spines of books on a shelf. I do this for two reasons. First,

the majority (though by no means all) of wild honey bee nests that I have seen – usually in cavities in trees – have positioned their comb the cold way, suggesting this might be what they prefer. Second, with the frames arranged this way it is possible to stand on either side of the hive during an inspection – easily being able to grab a lug in each hand and remove the frames so that they are face-on to you. With frames arranged the warm way, you can only stand behind the hive during inspections – or in front, which will annoy the bees. The location of your hives might affect whether you keep your bees the warm or cold way, and you can always turn the brood box around – keeping the entrance of the hive in the same direction – if you change your mind. With a rectangular hive like the Langstroth, frames will always go the cold way.

BELOW A hive with frames positioned in the cold way.

Essential equipment

When you first look at beekeeping supplies catalogues you will be baffled – or maybe excited – by the huge range of tools and gadgets that are available. In fact, few tools are needed, and usually the more basic they are the better. Buy the following equipment to start with, and acquire more only if you think you need it. It is useful to have a sturdy container to keep everything together. A stout canvas bag, plastic toolbox or even a bucket will do.

Protective clothing

A bee suit is a must. They can vary enormously in price – the best-quality ones costing several times more than the cheapest. Bear in mind that you will spend some time in your bee suit, so you need something that is durable, comfortable and well ventilated – it can get very hot in there. At first glance there may not appear to be much difference between the bargain suits and the high-end ones, but the more you pay the more comfortable it will be and the longer it will last. It is the zips that usually let down cheaper suits, particularly as they need to go through the washing machine frequently in beekeeping season.

You will need to decide whether you prefer the traditional all-around hat and veil face covering or the newer and more popular fencing veil style. This is a matter of personal preference. At beekeeping shows, the main suppliers will allow you to try before you buy – and often there are bargains to be had. It's said that bees are less threatened by a white suit, thinking that anything in dark colours is a potential threat, but you can buy a pink or lilac one if that's what catches your eye.

Bee suits have elasticated ankles which will fit tightly over shoes and boots, but the best way to protect your ankles (one of the more painful places to be stung) is to wear wellington boots and tuck the trousers in.

Gloves

Traditionally, beekeepers have worn thick leather gloves with long sleeves reaching up their arms. These are now very much out of favour; the smell of the leather can upset bees, and once stung the gloves retain the smell of the venom, which provokes further stinging. Leather gloves don't wash well and could harbour disease, and the thick material means you can't feel what you are doing with your fingers – making you more likely to squash bees and provoke defensive behaviour. Most beekeepers now wear disposable nitrile or latex gloves which allow them to better feel what they are doing and keeps their fingers free of sticky, staining propolis. By all means start off wearing thick leather gloves if it

ABOVE Suited and booted. The suit is a modern ventilated type with a fencing-veil hood.

ABOVE Nitrile gloves.

makes you feel safer, but graduate to the thinner ones as soon as you can. When buying disposable gloves, get those with longer cuffs that can be pulled up over the sleeves of your bee suit, preventing gaps that bees might crawl through. Wearing gauntlets (a separate sleeve with elasticated ends) to cover the join is a good idea. Buy tough disposable gloves that can be washed in a solution of washing soda (see page 57) between hives and after use. Reusing the gloves helps to reduce plastic waste, and biodegradable nitrile gloves are now also available.

Hive tool

Probably the most important piece of beekeeping equipment after the hive itself, a hive tool is kept in in your hand whenever you are working with bees. Hive tools are used for levering open hive boxes, scraping away propolis, loosening and lifting frames, and all manner of other jobs.

There are two basic designs with all sorts of variations and price tags. The standard hive tool has a flat, chisel-like blade at one end and a right-angle bend at the other end with another sharp edge. The 'J' tool has one flat, bladed end and one curved J-shaped end, from which it gets its name.

The flat, sharp end of each type is used to separate boxes and scrape away propolis. The bent or hooked ends are used for separating frames and, in the case of the J type, to lever the lugs upwards so that you can grab them with your fingers. Some beekeepers much prefer one type over another, while others use them inter-changeably. There is even a tool that combines all of the above features. You will find what works best for you.

Some hive tools have wooden handles or additional features. You don't need anything fancy – the best

cost only a few pounds. It is better to buy several of an affordable type so that you always have a spare as they are very easy to lose. Get used to holding a hive tool in your hand while doing all sorts of tasks in and around the hive. It should feel completely comfortable being in your hand all the time, ready to use whenever needed. If you put it down you are bound to lose it and then struggle to finish inspecting your bees properly. Using a bungee to attach a hive tool to your belt is one way of keeping it safe and at hand. Painting it with bright colours (dayglo nail varnish works well) or wrapping the handle with coloured tape makes it easy to find if you drop it in long grass – much better than finding it with your lawnmower.

Smoker

Smokers should be lit every time you open a hive, though not necessarily used. Their design is more or less the same whichever one you buy, but there are two key things to remember. A lit smoker is basically a bonfire in a tin can; the smaller the bonfire, the harder it is to light and keep lit for a long time. Even if you only have one hive, buy the biggest smoker you can afford. Because you have a fire going inside it, a smoker can get very hot; make sure you get one with a wire guard around the fire chamber to prevent burns.

ABOVE A good-quality smoker is a wise investment.

ABOVE Chisel- and j-type hive tools.

Other basic kit

Queen catcher

These are used for catching queens when you want to mark them with coloured paint or when removing them from the hive. The so-called crown of thorns type has a circle of pins embedded in a wood or metal circular mesh-filled frame. The cage is lowered over the queen and the pins pushed into the comb to trap her in place. She can then be marked through the mesh. The plunger type has a clear plastic tube with mesh at one end and a separate circle of sponge on a stick – the plunger. The plunger is used to guide the queen into the tube and keep her there. She can be gently pushed against the mesh to allow her thorax to be marked.

Queen cages

These are matchbox-sized cages of various designs, used to keep the queen safe during beekeeping operations or when she is being introduced to a new colony.

Blowtorch or matches

These are used for lighting your smoker. Using a self-igniting gas blowtorch is the fastest way to light a smoker, but matches do the job. It's a good idea to keep a spare box of matches wrapped in a plastic bag for when your other matches get damp, or the blowtorch runs out of gas.

Marking pen

This is a water-based paint in a pen. Young queens are marked with a coloured dot on their thorax. The dot makes queens easier to spot in the hive, and a different colour is used each year so you can always tell what year a queen was born. If you keep good records, you should always know how old your queen is, in which case you can always use the same colour every year. White is the best colour for making a queen easy to spot.

Drawing pins

Marking frames with drawing pins can show which need to be replaced or where a queen cell is located. They can be used for attaching mouseguards (see page 149) or marking the outside of a hive as a reminder of something important. Keep a few pinned to the inside of each hive roof and always carry extra ones.

Uncapping tool

Originally designed for uncapping honey cells before harvest, this tool is now primarily used to check drone pupae for the presence of varroa (see page 143). When out of use, keep the prongs embedded in a block of cork or polystyrene, as they are needle-sharp.

Magnifying glass

Useful for studying comb for eggs, brood or signs of disease. The type with built-in LED lights is particularly useful.

Frame rest

When inspecting a hive, it is useful if one or two frames are removed, making room inside the brood box for other frames to be loosened and removed. A frame rest is hung on the outside of a hive and one or two removed frames hung temporarily within it.

Bucket and lid

A small, sealable bucket or container can be used during hive inspections to collect pieces of removed wax. Wax left on the ground can attract wax moths. Every scrap of wax that you save is useful for making candles or other wax products – it's surprising how much you can collect.

Washing soda crystals

The main cause of disease spreading from one colony to another is likely to be the beekeeper. If you have more than one colony, you should clean your hive tool before moving from one to the next. Keep a plastic container with a tight-fitting lid half filled with a strong solution of water and washing soda with an added squirt of washing-up liquid. Keep a wire wool scrubbing pad in the container too. After inspecting each hive, dip your hive tool in the solution and use the wire wool to scrub it clean. While still wearing them, and being careful not to rip them, wash your latex gloves in the soda solution as well.

Washing soda is capable of killing the pathogens that cause some bee diseases. It is also one of the only things that will dissolve propolis, removing it from tools and gloves. It is an organic substance (sodium carbonate) and it is fine to pour small quantities of the mixture down the drain when it begins to become ineffective at cleaning.

Soda crystals can be added to the soap powder used when washing your bee suit in the washing machine. This will remove germs and propolis from the suit, and should be done regularly during the summer and certainly before and after visiting other people's bees. Veils should be hand-washed separately, following the manufacturer's instructions.

Keeping bees

The best time to acquire bees is from mid-spring to early summer. At this time the main summer nectar flow is still to come, giving a small, young colony – called a nucleus colony – the chance to grow in size and strength before winter. Over the summer, you can practice regular beekeeping tasks and develop your bee-handling skills, gaining confidence as the number of bees in your colony grows. If you start with a queen raised that season, the colony is unlikely to try to swarm until the following year. This gives you plenty of time to prepare your swarm management plans.

What kind of bee?

Although the western honey bee, *Apis mellifera*, is a single species, over thousands of years distinct subspecies have evolved to survive in particular geographical regions with their varying climates, seasons and flora. For example, the Italian honey bee, *A. m. ligustica*, is subtly different from the Spanish honey bee, *A. m. iberica*.

The native British honey bee subspecies, *A. m. mellifera*, is known as the black or dark bee and is perfectly adapted for life on these islands. Nevertheless, for more than 150 years British beekeepers have imported huge numbers of queen bees from abroad (more than 21,000 in 2020) in the belief that other races might produce better crops of honey or be more resistant to disease or easier to handle. As with other forms of livestock, there have even been selective breeding programmes to produce a hybrid 'superbee' that combines the best aspects of each subspecies. This type of bee is known as a Buckfast, after the Devon abbey where such breeding techniques were pioneered by a Benedictine monk called Brother Adam. Confusingly, Buckfast refers more to an approach to breeding than to a distinct type of bee, which means that if you buy a Buckfast queen you don't really know what it is you are getting. Over the decades, imported bees have interbred naturally with native bees, meaning that most UK bees today have a complex genetic heritage and can be highly variable in terms of seasonal behaviour, resistance to disease, productivity and temperament.

Many amateur beekeepers now believe that importing queens is unsustainable. This is partly because of the potentially destabilising effect it might have on the genetic makeup of existing bee populations, but also because it risks the accidental introduction of non-native pests or diseases. The prevailing trend among hobby beekeepers is to aim for what are known as 'local' bees. These are bees that have lived in a relatively small geographic area for many years, the beekeeper breeding from only the best-performing colonies. The result is an ecotype – a bee that, whatever its genetic heritage, is highly suited to the local environment and should be good-tempered and productive.

OPPOSITE PAGE Forager on lavender.

RIGHT The apiary at Buckfast Abbey, Devon.

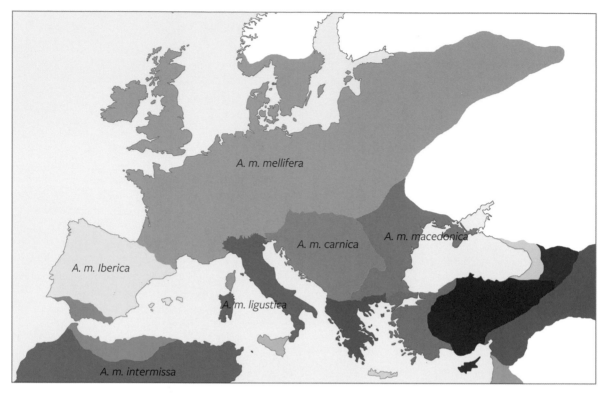

ABOVE Around 30 subspecies of *Apis mellifera* originate from distinct geographical areas of Europe, Africa, Asia and the Middle East. Each coloured area on this map indicates the natural range of a subspecies, with some of the more significant being named.

ABOVE A well-populated six-frame nuc of bees.

Local bees can usually be sourced via local beekeeping associations, whose members might raise and sell small numbers of queens and nucleus colonies from their best stock. If you cannot find locally bred bees, try one of the many commercial bee suppliers. These may sell both imported queens and home-bred queens, so make sure you know what is being offered before you make a purchase, and buy UK-bred queens if at all possible.

Acquiring bees

It is possible to start beekeeping with a swarm (see page 63), but most new beekeepers start by acquiring a nucleus colony, also known as a nuc. Nucs are composed of five or six frames of bees and a queen, housed in a half-sized beehive called a nuc box. The frames should comprise fully drawn wax comb, with approximately half of them containing food and half containing brood in all stages (eggs, larvae, sealed brood). The queen could have been produced towards the end of the previous year, in which case the colony will have been overwintered. Alternatively, she might have been produced in the early part of the season in which you are buying your nuc – the exact time of availability being highly dependent on the weather. Overwintered nucs are always in high demand for new beekeepers keen to get an early start, and so are usually highly priced. Nucs bought a month or two later are usually cheaper. A nucleus colony should expand very rapidly and will be ready to transfer into a full-sized hive within a few weeks of purchase, if not immediately.

Bringing bees home

Nucleus colonies normally have to be collected from the supplier. This means having to transport them in your car, and this needs careful consideration. The greatest concern is that the bees might overheat if subject to a long journey on a hot day, or the wax comb could soften and collapse. Under normal conditions a car journey of

an hour or so is fine, but precautions must be taken if it is likely to be more than that on a warm day. The solid roof of the nuc box can be replaced with a travelling screen – a temporary roof made from bee-proof mesh. A damp sponge placed on top of the mesh will provide water and help to keep the bees cool, or the mesh can be lightly sprayed with water now and then. Travel with the car windows open to provide a flow of air. The travelling screen or roof should be tightly strapped to the nuc to prevent separation. It's a good idea to drive wearing your bee suit with the hood down, ready to use should bees escape during the journey. Needless to say, the door of the nuc should be very firmly closed. Position the nuc so that the frames in the box point in the direction of travel. This will limit how much they swing back and forth, which could squash bees or even your precious queen. Wedge the nuc in place with boxes or cushions so that it won't move or fall over should you swerve or brake suddenly. Ask the supplier to check your travel arrangements before you leave.

OTHER POINTS TO LOOK OUT FOR WHEN BUYING A NUCLEUS COLONY

- The bees should be of good temperament when handled appropriately, and the supplier should be prepared to let you handle and inspect them before purchase.

- Make sure the nuc colony is on the correct size and type of frame for the hive you have bought.

- The wooden frames and comb should be new and of good quality.

- All frames should be well covered with bees and have a balance of young/house bees and older foraging bees.

- The queen should be correctly marked to indicate the year of her birth. The supplier should be prepared to clip the queen's wings if requested (see page 86).

- The brood should all be the offspring of the queen in the nuc.

- There should be no sign of any queen cells (which indicate that the colony wants to swarm or replace their queen – see page 81).

- There should be no signs of disease in adult bees or brood. Varroa will almost certainly be in any nuc that you buy, but varroa or varroa-related disease should not be noticeable.

- The supplier should be able to provide a complete record of any treatments used. It is a legal requirement to record such treatments and pass them on to the buyer.

- Any treatments used should be approved by the Veterinary Medicines Directorate as listed on BeeBase, the website of the National Bee Unit (see page 297).

- The nuc box could be a well-constructed wooden box that will last many years, or be made from plywood, Correx or even cardboard. Most nucs now come in polystyrene nuc boxes which are particularly good at insulating small, young colonies. A good-quality wood or polystyrene nuc box is an essential piece of equipment that you will re-use time and again, so it is worth acquiring one from the start. When considering advertisements for nucleus colonies, make sure you know what kind of nuc box is included in the price.

Placing and rehoming your bees

When you get home, your chosen apiary site should be prepared with a hive stand in its final position. Place the nucleus box on the stand with the entrance pointing in the right direction, replace the roof if necessary, and then open the door to release the bees. Sometimes bees flood out in a state of agitation (have that hood ready!). Other times, they emerge cautiously. Over the next few hours, you will see bees leaving the hive and flying around it in ever increasing circles. These are orientation flights – the bees are learning the location of their new home.

If your nucleus is very full of bees and brood it will need to be transferred to a hive within a few days of arrival. At most, this should be done within a few weeks. Ask the seller for their opinion about how soon it should be done. If you leave it too long, it could check the growth of the colony, or your bees might swarm.

To move the bees from the nuc box to a full-sized hive, wait for a warm day (a minimum of 14°C) and have the hive and all of its parts ready, as well as an additional five or six brood frames fitted with wax foundation. Light your smoker (see page 69) and give one or two gentle puffs of smoke around the entrance of the nuc. A small nucleus colony is likely to be calm and easy to handle without the use of smoke, but at this stage of your beekeeping career a little smoke, gently used, is no bad thing. Wearing your bee suit, gently lift the nucleus box off the stand and place it on the ground to one side. Then place the floor of the full-sized hive on the stand, followed by the brood box. Next, use your hive tool to loosen and separate the frames in the nucleus hive. If there are lots of bees on the top bars of the frames and it is hard to use your hive tool or to get your fingers around the frame lugs, a whisper of smoke over the top will send the bees down onto the comb and out of your way. Slowly and gently lift out the first, outer frame and transfer it over into the brood box of the new hive. Do this very carefully so that no bees are knocked off the frame, in particular the queen, whose safe transfer is vital. If you are confident and very careful you can check each side of the frame to see if the queen is there – it is comforting to know she has made it safely into her new home.

Transfer each of the combs from the nuc box to the hive, making sure they go back in the same order and that the combs face each other in the same way as before. If you are keeping your bees the warm way (see page 53), place the frames at the front of the hive. If you are keeping your bees the cold way, place the frames in the centre of the hive with a gap on either side. When all of the frames are transferred, fill the remaining space in the hive with new frames of foundation, gently pushing all the frames together to close any gaps. Finish by adding a dummy board. Now check how many bees remain in the emptied nuc, as there are usually a few clinging to the walls and floor. Inspect them to make sure the queen is not among them. If she is, very gently brush her into the new hive. Then turn the nuc box upside down over the hive and give it a firm shake so that any remaining bees fall into their new home. Another light puff of smoke will send bees down between the frames, and you can add the crownboard followed by the hive roof.

You have successfully transferred your nucleus colony into your hive. The bees will barely notice the difference and will carry on leaving and arriving through their new front door – which will be in exactly the same position as their old one. The colony now has space to fully expand over the following months, although if the weather is poor, you may wish to feed them (see page 129).

BELOW A polystyrene nuc with wooden stand and roof.

Collecting a swarm

Collecting a swarm is another way to acquire your first colony of bees. Most beekeeping associations have a list of experienced beekeepers who are prepared to collect swarms, and a list of new beekeepers hoping to receive one. If your name is on the list, you might be contacted by a swarm collector and invited to help them collect a swarm, or they might arrive on your doorstep at short notice with a box full of bees ready to go into your empty hive.

As a new beekeeper, there are advantages to starting with a swarm. The bees cost you nothing and, once installed in a hive that they approve of, a strong swarm is astonishingly quick to build comb and fill it with brood and stores. But there are disadvantages too. You might wait all summer without getting a swarm, a late swarm may not have time to grow sufficiently before winter, you won't know if the bees are diseased or of a bad temperament, and you won't always know the age and condition of the queen. Despite that, most swarms are problem-free.

Swarm collecting is an essential beekeeping skill and, once you have learned the basics, you will find that nearly every swarm requires a slightly different approach. Sooner or later one of your colonies is likely to swarm and you will want to catch it and put it back in a hive. Try to assist with one or two swarm collections before attempting it on your own.

When a colony of honey bees swarms, about half the bees and a queen leave the old nest and fly in a swirling cloud until they find somewhere nearby to settle – usually on a branch, but perhaps on a fencepost, in a hedge or even on a parked car. They come to rest usually 10–20 metres from the site of their original nest, and form a cluster like a large bunch of grapes. Scout bees then fly out to search for somewhere new to live. Only when there is agreement about the suitability of a new home will the cluster break up and the bees fly away. This can happen anything from half an hour to several days after swarming. It is while the bees are in the initial clustering stage that they can be captured and re-housed – a job that sounds thrillingly dangerous but which is usually quite straightforward, although sometimes a little challenging, and always very enjoyable.

Swarms are usually remarkably placid, but it is a good idea to wear a veil and gloves while working closely with them. The exception can be when a swarm has been unable to find a new home for a day or two, in which case the bees may be hungry, frustrated and less welcoming. A light spray of sugar water will usually help to pacify them. The most dangerous thing about swarms is the risk of falling off ladders or into hedges when trying to reach them. Be realistic about the potential dangers, especially when heights are involved. If you do use a ladder, someone (in a bee suit) should be there to hold it steady.

A skep is the traditional container for collecting swarms, but the captured bees will still need to be transferred into an empty hive where they can establish a

WHAT YOU WILL NEED

- An empty nuc box and set of frames and foundation
- A bungee cord or ratchet strap
- A pair of secateurs or loppers (ideally both)
- A bee brush – a bunch of long grass is fine
- A water spray bottle containing weak sugar solution
- Smoker and materials
- Bee suit and gloves
- A stepladder

BELOW A swarm collected in a skep, with some junior assistance.

ABOVE LEFT AND RIGHT Conveniently located swarms can be knocked straight into a nuc box. One or two frames of foundation give the bees something to cling on to.

RIGHT Closing up the nuc when all the bees are inside.

new home. A much easier method is to use a polystyrene nucleus box. These are light enough to lift up to wherever a swarm is hanging and, once inside the box, the bees only need to be given some frames of foundation before the roof is put on and the captured colony can be taken away. They will quickly build comb and can easily be transferred to a full-size hive when they have outgrown the nuc box.

When collecting a swarm, the aim is to quickly and efficiently get as many of the bees into a container as you can. The one you really need is the queen, after which the others will follow. If the swarm is hanging conveniently from a branch, try to get into a position below the swarm so that you can lift your nuc box upwards until the whole cluster, still attached to the branch, is hanging inside. You might have to carefully cut away surrounding leaves and twigs to do this. When the swarm is inside, very firmly shake the branch to dislodge the bees. Go hard and fast with this – two or three really good shakes should be enough. It is useful to have an assistant so that one of you can hold the nuc box while the other does the shaking. Many bees will fly into the air, but most will land in a heap on the floor of the box. Quickly brush any bees remaining on the branch into the container.

As quickly as you can, take the container to ground level and give it a sharp knock so that bees climbing the sides to make their way out fall back to the bottom. Carefully lower a full set of frames and foundation into the box and then replace the roof and open the entrance hole. Now, watch what happens. Many bees will leave the nuc box and fly off, but within a minute or so you should see worker bees arranging themselves around the entrance, raising their abdomens and fanning their wings. These bees are wafting lemony-smelling Nasonov pheromone into the air as a homing beacon. If this happens, it means that the queen is probably inside, and the signal to come and join her is being sent out. Gradually, the flying bees will follow the scent to the

ABOVE A worker lifting her abdomen and fanning out the Nasonov 'homing' pheromone.

the container. If the bees are sprawled on a hedge or clinging to a post, rest the container over the top and waft a little smoke under the bees to encourage them to move up into the darkness of the container. If the swarm is on a flat surface like a window or the side of a car, study the swarm and look for the queen. If you find her, put her in a queen cage and then carefully scoop up the bees using a sheet of cardboard or even a dustpan, and tip them into a container with the caged queen. You will be amazed at the different ways in which you will catch swarms over the years, and the interesting places you get to visit when doing so.

Who owns a swarm?

If you are called to collect a swarm, how do you know whose bees they are and whether it is your right to collect and remove them? There is a simple rule here – finders keepers. Beekeepers are responsible for their bees, and this includes preventing them from swarming. Once bees have left someone else's hive or property, they are legally considered *ferae naturae*, or wild animals. If you think you know who a swarm belongs to, it is good manners to let them know that you have found it and ask if they are going to collect it. However, it can be very hard to tell where a swarm hanging in a tree has come from.

You must ask for permission from the landowner if you wish to remove a swarm. This includes permission for access and possibly to make a path through vegetation

nuc box and go inside. If a few bees cluster back on the branch, brush them off again. If the cluster entirely re-forms on the branch, it means you haven't got the queen and will need to try again.

To remove the captured swarm, try to wait until dusk when all flying bees should be settled. It's fine to go away and leave it unattended for a few hours, but place a warning sign on it if there is any possibility that members of the public might come near it. Close the door of the nuc and secure the lid in place with a bungee cord or ratchet strap to prevent any accidental escapes. If you are transporting the swarm in a car, wedge the nuc box securely so that it doesn't fall over while in transit. It is a good idea to quarantine swarms at a location at least 3 miles (5km) from your apiary, and once there is capped brood check for signs of brood disease (see page 139) before the young colony is moved nearer to your bees. For the first week or so after collecting a swarm, leave a queen excluder across the entrance (these are built into the circular entrances of most nuc boxes) so that the queen can't escape and take the swarm with her – a behaviour known as absconding. Once the queen is laying in the nuc box and there is brood, the bees will consider it home and will not leave.

There are many variables when catching swarms. If the swarm is on a thin branch, you can gently cut away the whole branch and lower it and the swarm into

BELOW Some swarms can present a challenge.

or remove branches if necessary. Swarm-collecting is a quid pro quo arrangement; the landowner should be glad to get rid of the bees, the beekeeper should be glad to obtain them.

You should never charge to remove a swarm because that can make you liable for any damage caused to property. However, there is nothing to say you shouldn't accept a cup of tea and a piece of cake for your efforts.

Bee stings

Getting stung is inevitable when you keep bees – but it should happen only rarely. If you get stung a lot, it's possibly because you are doing something wrong.

Honey bee workers sting to defend their nest, only doing so if they feel threatened by a potential predator. Away from the nest, when visiting flowers for example, bees are very unlikely to sting. Defensiveness in bees can be a genetic characteristic, and if your bees seem to sting for no apparent reason, or do so at some distance from their hive, it might be that they are genetically more inclined to do so.

Any of the commercially available subspecies of bees such as Italians (*Apis mellifera ligustica*), or the hybrid strain known as the Buckfast, should at first be easy to handle and very unlikely to sting. However, once they produce their own replacement queens, which then mate with drones from other nearby colonies, the genetic makeup of your bees will change. The result of this genetic lottery can be that previously gentle bees become more defensive – possibly even downright unpleasant. This does not always happen by any means, but when it does it can be enough to make some people reconsider keeping bees.

One way to make sure your bees remain good-tempered is to replace the queen every two or three years with a new queen of the same race. However, this traps you in an expensive and demanding cycle of buying new queens from some distant supplier. Furthermore, you will continually be introducing novel genes to the local bee population, possibly exacerbating the problems caused by the mixing of strains. This is one reason why many beekeepers now prefer to use only bees bought from local beekeepers known to have good-tempered colonies.

There are times when even the most mild-mannered bees are more likely to sting. Chemicals used by farmers or gardeners can sometimes cause bees to become bad-tempered, as can a recent visit to the hive by a potential predator such as a dog or a badger. Bees that have been enjoying a bounteous nectar flow can become grumpy for a day or two when it suddenly stops. In particular, this can happen in rural areas when a local oilseed rape crop has finished flowering. Using noisy garden machinery near hives can put bees on edge, as can close, thundery weather. A sudden rise in defensiveness might also occur because a colony has lost its queen. If normally passive bees become a little bothersome, avoid inspecting them for a while, or wear a veil if you have to work near them in the garden. They usually calm down in a day or two.

Most stings happen when beekeepers open their hives – hardly surprising since this involves dismantling and disturbing their nest.

The secret to working safely with bees is being gentle and slow in everything you do. Follow these guidelines when working with bees and you will be less likely to get stung:

- Make sure gloves, veils and bee suits are all tightly sealed – most stings happen when curious bees find their way into small openings and get trapped.

- Wear clean gloves and clothes – lingering odours can provoke more stings if you were stung during your last visit.

- When moving near the hive, try to keep to the back and the sides – away from the defended entrance.

- Smoke bees gently and sparingly. Allow initial smoke to take effect before opening the hive.

- Use a hive tool to gently loosen hive parts and frames before removing them, thereby avoiding unnecessary jarring.

- Avoid banging or jolting hive parts when inspecting bees.

- Limit how much you bend over the hive or move your arms back and forth over the frames. Such overhead movements can be perceived as threatening.

- Do not swipe bees away if they fly around or land on you.

- Avoid squashing bees with your fingers or equipment.

- Monitor the mood of the hive. Use a little more smoke if the bees become agitated, or close the hive if it feels like things are getting out of control.

- Keep inspection times to a minimum.

- Inspect hives during the warmest part of the day, when most foragers have left the hive.

- Some perfumes, hairsprays and other toiletries can irritate bees. Don't wear any of these products if the bees seem to object.

Getting stung is never pleasant, but stings can vary in intensity, and reactions may differ. In most cases the result is immediate sharp pain, followed by swelling or a rash. The pain may bother you for a few minutes or a few hours. Most beekeepers get used to being stung and treat it as an inconvenience, rather like being stung by nettles. A tiny number of people are allergic to bee stings and, very rarely, getting stung can lead to death. If you start to feel dizzy, tight-chested or panicky after being stung you should seek immediate help, and call an ambulance.

If you have nearby neighbours, their safety should be your priority. They might get stung at some point and you will need to discuss this possibility with them before it happens. Explain that stinging is rare and that you will do your best to minimise its likelihood. Some people are surprisingly blasé about being stung, but for others the very idea is intolerable. Being generous with your honey harvest can be surprisingly effective at encouraging neighbourly understanding and forgiveness.

To minimise the risk of neighbours getting stung, set up your hives as described earlier to encourage your bees to fly high over neighbouring property. Try to only open and inspect your hives when you know your neighbours aren't in their garden. Keep only good-tempered bees, and if a colony becomes bad-tempered, move it away until it can be brought under control, usually through re-queening.

Other pets

Honey bees present few dangers to other pets or animals. Inquisitive dogs might get stung if they poke their nose into a hive entrance, but they usually only do it once. If your hives are in a field, make sure they are properly fenced so that livestock don't use them as scratching posts, or they will end up with more than just an itch. It's often said bees that don't like horses, but in most circumstances there is nothing to worry about as long as they are not very close neighbours.

BELOW Old-fashioned leather gloves, which should be avoided.

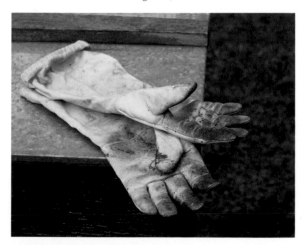

WHAT TO DO IF YOU GET STUNG

- Remove the sting as quickly as possible by scraping it away with a fingernail or hive tool. Do not grab the protruding part of the sting with your fingers as this will squeeze more venom into the wound.

- Move away from the hive.

- Puff some smoke around the sting to mask its smell – the smell of the venom can encourage more bees to sting.

- Take a minute to check that you are feeling well and are not experiencing any of the more serious symptoms mentioned earlier.

- Apply a sting relief product if necessary.

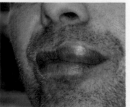

Handling bees

Smoking bees

As a species that evolved to live mostly in trees, honey bees seem to have an instinctive reaction to smoke; it perhaps signals that a fire is nearby and that their nest could be under threat. In response, the bees gorge on honey, preparing to take the precious but portable resource with them should they need to abandon their home. The rescued honey will fuel an escape flight and can be used in the production of wax to quickly establish a new nest. This emergency preparation keeps bees busy and makes them less likely to defend their nest against predators – or beekeepers. It also makes it harder to mount a defence as bees with a tummy full of honey find it difficult to bend their abdomen and deploy their sting. Furthermore, threatened bees emit alarm pheromones that instruct other members of the colony to take defensive action; smoke disguises these signals and disrupts a colony's call to arms. Beekeepers have long taken advantage of these natural effects, employing smoke to help to subdue and manage colonies when working with them.

Smokers

Smokers are very simple pieces of equipment and there is not much difference between one design and another. The two main points to consider when purchasing are size and safety. As mentioned earlier, it is best to buy the biggest one that you can use comfortably. It is much easier to make a long-lasting fire in a large smoker, but you also need to be able to handle it with some dexterity. To avoid accidental burns, make sure your smoker has a wire guard around the fire box, which can get searingly hot.

OPPOSITE PAGE A swarm being encouraged to move up into a skep.

Lighting a smoker

Lighting and maintaining a tiny bonfire inside a small container is surprisingly tricky until you get the hang of it. The key is to choose the right fuel and to get it burning well from the start. The aim is to produce a continuous, gentle trickle of smoke which will increase to a cloud of cool, white smoke after a couple of puffs on the bellows. Once lit, a smoker should remain active for about half an hour and can be kept going longer with the occasional addition of extra fuel.

All beekeepers have their favourite smoker fuels, and you will find what suits you. What is needed is something that burns slowly and produces plenty of pleasant, white smoke. Wood shavings, rotten wood, dried lavender, hay and hessian sacking are all commonly used. Whatever you choose, make sure that it hasn't been treated with any kind of chemical that might harm the bees. Do not wear the veil of your bee suit when lighting a smoker as sparks can melt the mesh or even set fire to it.

Using a smoker

Smoking bees is a skill that can take some time to master. Not all colonies need to be smoked, and some need more than others. All will react a little differently each time you pay a visit. Even if a colony is usually very gentle and you feel it shouldn't need smoking, a lit smoker should always be at the ready. A smoker is not a weapon to be used in your own defence. By giving bees a little smoke before you first disturb them, and a little more now and then as your inspections progress, you can keep most colonies calm and easy to work with. If bees become defensive, puffing smoke at them will only make them more agitated. It is better to replace the roof and come back later.

Smoke does not take effect immediately, so is best administered a little before you open a hive. Give the bellows of the smoker a couple of vigorous puffs to

1. Use newspaper to start your smoker. Crumple up a fist-sized ball and light it with a match or gas torch. Use your hive tool to push it down to the bottom of the firebox, giving a light puff on the bellows to keep it burning.

2. Slowly add a small amount of your chosen smoker fuel, puffing on the bellows to get the fire going.

3. Leave the smoker burning with the lid open while you make your other preparations. Give the bellows a squeeze if the fire needs some help.

4. When the fire is hot and there are embers in the firebox, add a deep layer of your fuel. This should produce thick smoke but no flames. It might die down to a very small trickle of smoke, but should produce much more with a puff or two of the bellows.

5. To produce cooler smoke, add a twist of green grass on top of your fuel. This will also stop any hot sparks coming out with the smoke.

6. Squeeze the bellows gently to puff small amounts of smoke into the hive when needed.

get some nice white smoke going, and then administer several gentle puffs near the hive entrance, allowing the smoke to drift into the hive – don't blast it in. Give the colony a couple of minutes to settle down before removing the roof. Use your hive tool to lift the crownboard slightly, and gently puff a small amount of smoke into the hive. Using smoke in this way takes advantage of its ability to direct bees – a puff of smoke at the top of the hive will send bees down between the frames and out of your way. The ability to direct bees like this them is one of smoke's most useful attributes.

Smoke might be used every five minutes or so during a hive inspection, depending on the mood of the bees. Although it's easy to get engrossed in whatever else you are seeing and doing, continually observe how the bees are behaving and what they sound like. With a little experience you will detect the early signs of increasing agitation, including buzzing at a slightly raised pitch, and then you can use a little smoke to calm things down before they become more difficult to deal with. Even if you don't use your smoker, remember to puff the bellows every so often so that the smoker stays alight and ready should you need it.

At the end of an inspection, smoke is again useful for directing bees. A waft of smoke around the edges of boxes and over the tops of frames will move bees out of the way so that a hive can be put back together without any bees getting squashed.

After use, block the nose of the smoker with a cork, a trimmed stick or a handful of grass. Deprived of air, it will go out and slowly cool down. Never store a smoker until you are sure it is completely extinguished – beekeepers have lost sheds and even cars through smoker negligence. Fireproof metal smoker storage boxes are available.

Looking at bees

The most important beekeeping skill is observation. You will spend more time looking at your bees than on almost any other beekeeping task. Making careful observations and relating these to what you know about honey bee behaviour and biology will help you to make decisions about managing your bees.

In winter and spring, when it's too cold to open the hives and see inside, observations might include how frequently the bees fly from the entrance, whether they are bringing in pollen, if there are any dead bees outside the hive and whether nibbled wax cappings from emerging brood are dropping onto the inspection tray. After a while, making observations like these and drawing conclusions from them will become second nature.

From late spring until early autumn, you can open hives and see what is happening inside. Such visits are called hive inspections. You will inspect your colonies infrequently at the beginning and end of the beekeeping season, but in early summer and mid-summer it usually needs to be done weekly, mainly to manage swarming. As well as checking on the condition of the bees, various housekeeping matters usually need to be attended to, such as replacing old, bad-quality comb with new frames of foundation.

As a beginner, take every opportunity you can to observe experienced beekeepers working with their bees; they will talk you through what they are seeing and doing and you will get used to the various inspection procedures and techniques. These can be the most valuable learning experiences of all.

Making inspections

Bees are best inspected on warm days with little wind and a temperature of 16°C or above. The best time is usually mid-morning to late afternoon, when the bees are flying and many are away from the hive.

Gather all of the necessary equipment, making sure there is nothing you might have to dash home for. Wear your beekeeping clothing, only putting up the veil just before a hive is opened. Light your smoker before you put up your veil, and make sure it is going well. Give several puffs of smoke around the hive entrance and follow the smoking guidelines outlined earlier. While you wait for the smoke to take effect, it is a good time to reflect on what you need to look for when the hive is open. A checklist is useful – these are usually incorporated in a record card (see pages 78–79). The following are important to note:

- Does the colony have a queen and is she laying eggs?
- Is the pattern of eggs and brood appropriate for the time in the season?
- Is there sufficient room for the queen to lay more eggs and for the colony to expand?

- Is there sufficient room to store nectar and honey in any expected flow?
- Are there enough stores of honey to last the colony until the next inspection?
- Are there any signs of queen cells, and what stage are they?
- Are there any signs of disease, including varroa?
- Do any damaged or poor-quality frames need replacing?

Because there are so many things to remember, it can be easy to find yourself rushing through hive inspections. Try to take your time, as the bees won't thank you if you do it in a hurry and are clumsy or rough with them as a result. There is sometimes a definite limit to how long bees are prepared to tolerate your meddling, and they will let you know when it is time to close the hive – but on warm summer days it is surprising how long you can have a hive open. This is your chance to carefully observe the world of the colony and watch closely the behaviour of individual bees as they work and interact with one another. What you see will always be utterly fascinating, so take some time to appreciate what is one of the most absorbing and rewarding aspects of beekeeping.

Opening the hive

First, pick up your hive tool and hold it firmly in your dominant hand. It should stay there throughout the inspection process. Stand behind or to one side of the hive depending on whether the frames are arranged in the cold way or the warm way. Remove the hive roof, being careful not to jolt or bang it against the hive – working delicately throughout will help keep your bees in a good mood. Place the roof upside-down on the ground next to the hive. You will stack other hive parts on top of it. Insert the blade of your hive tool between the crownboard and the box below and lever up the crownboard just enough to puff in a little smoke. Remove the crownboard – you might need to insert the hive tool and prise upwards in several places to break the propolis seal. If the crownboard is directly over the brood box, carefully check the underside to make sure the queen is not on it – she is unlikely to be here, but gently brush her into the hive if she is. Place the crownboard on the ground, resting at an angle against a front corner of the hive.

ABOVE A puff of smoke under the roof will send the bees down.

ABOVE Remove the roof, avoiding knocks and bumps.

ABOVE Prise off the crownboard, and add another puff of smoke.

ABOVE Lift the crownboard and repeat procedure to remove super.

ABOVE Stack upturned roof and boxes nearby.

ABOVE Check each frame for brood. This one contains only stores.

If the hive has supers, these should be removed next. You will be able to gauge how much honey a super contains from its weight as you lift it. If you want to inspect individual super frames to see what the honey harvest is looking like, this is best done at the end of the inspection process, when the supers have been returned to the hive. Each removed super is stacked at a different angle on top of the upturned roof. This way each box touches the one below only at four small points of contact, minimising the likelihood of squashing bees. Once supers are removed, the crownboard can be placed on top to discourage the bees inside from flying back to the hive. This also lowers the chance of bees from other colonies finding the exposed honey and possibly trying to rob it.

If you have removed supers, the queen excluder will need to come off next. By now the bees in the brood box might be getting a little agitated, so a gentle puff of smoke over the queen excluder will calm them and send them down into the brood box. Remove the queen excluder and add it to your stack of equipment. If there are any bees on the excluder, make sure that none of them is the queen.

Inspecting the brood nest

You will now be able to see the top bars of the frames in the brood box. So long as you haven't used too much smoke you should get a sense of the shape of the brood nest. There will be a concentration of bees on top and between the frames where there is brood, with fewer bees on the frames containing only stores of honey and pollen. Usually, the overall impression is of an oval shape, with the bees more densely clustered around the brood, which they are feeding and keeping warm.

The main aim of the inspection is to check the brood nest, although you also need to look at the frames on either side to gauge the quantity of stored food. First, remove the dummy board if there is one. Hook the curved end of your hive tool under the lug at one end of the dummy board and lever it up. Dummy boards and frames are usually stuck together with propolis, so prise them apart gently. If you are too forceful this can snap off the frame lugs. When one lug is loosened, move to the other end of the dummy board and repeat. Then lever up the lug and grasp it in the fingers of one hand. Repeat at the other end to raise and grasp the lug there. Holding the dummy board at both ends, lift it out of the

ABOVE Carefully remove one frame at a time.

You now have a small gap on one side of the brood box which can be used as a working space. Next, prepare several brood frames for removal by first loosening them. Working on one side of the brood box at a time to minimise how often your hands pass back and forth over the hive, use the hive tool to lever the first frame away from the one next to it, pulling it over into the gap left by the dummy board. Then do the same for the next frame, and so on until several frames have been separated. Next, move your hand to the other side of the box and loosen the other ends of the frames. You will now have several unstuck frames with a small space between each of them.

During hive inspections, hold brood frames over the hive so that the queen will drop back into the hive should she fall off. Ideally, the sun should be behind and slightly to one side of you, making it easier for you to see into cells when inspecting frames. You may have to move or twist the frame or your body to get the best angle for viewing.

Frames are best kept more or less vertical during inspections. If a frame is held horizontally for prolonged periods, nectar can drip out, bees or the queen can fall off and, on a warm day, the wax comb can warp or even collapse.

hive. Raise it slowly and keep it level so that any bees on it are not squashed on the way out. If there are lots of bees on the removed dummy board, hold it over the hive and give it one firm downward shake. The bees will fall harmlessly back into the hive. Put the dummy board on top of the equipment pile.

Removing the brood frames

Lift the first loosened frame out of the hive, bringing it to eye level. Keep it vertical and avoid scraping it against anything to prevent damaging the comb or squashing any bees. It is likely to be a frame of stores, although early in the year it could be empty comb or foundation. When

ABOVE A j-tool can be used to lever up tight frames.

ABOVE A nice frame of brood.

Larva close to pupation

Pollen

Egg

Two-day old larva

Newly emerged larva

Prepared cell ready for egg

you have noted the content of the frame on both sides, check very carefully that the queen is not on it, and then put it to one side. It can go in a spare hive or nucleus box or on a handy detachable frame rest that hangs over the side of the hive. The idea is to store it safely while creating even more space in the hive (see page 57).

Remove the next frame in the same way. By now, inquisitive bees might be starting to come up to the top of the box and cover the lugs. You can wiggle your fingers gently past them, nudging them out of the way, or waft a little smoke over them to make them go down again. The second frame might contain more honey, or if you are getting closer to the brood nest, might contain more pollen. It might even contain eggs and larvae, showing that the brood nest is expanding in this direction.

Keep an eye out for the queen – though it isn't strictly necessary to see her. If there are eggs in cells then you will know that she must have been there within the last three days. The main reason for finding the queen is so that you can take extra care with the frame she is on, to avoid harming her or losing her.

When you have finished looking at the frame, replace it in the gap it came from, making sure that it goes back the right way around and is tight against the last frame that you looked at. It is important to make sure frames always go back in the same way because the surface of their combs can undulate, with each frame interlocking with the next to preserve bee space between the combs.

The brood nest

Several frames in (although it can begin at the first frame) you should encounter the brood nest. Usually, you will first find a small patch of eggs and young larvae surrounded by stores, but on subsequent frames the patch should get bigger and there will be increasingly large areas of sealed brood. As you progress through the frames the brood area will reach its maximum size and then begin to shrink again. The overall shape of the brood nest if looked at in cross-section is typically oval.

Each comb in the brood nest will contain cells that house the different component elements that are necessary for raising new bees. There is usually a distinct pattern to these elements, but the exact ratios will vary depending on the time in the season and the stage of the brood cycle on each frame. Typically, there will be sealed worker brood in the centre of a frame – where the first eggs were laid. Surrounding this there will be cells containing some large larvae close to pupation. Further out there may be some smaller, younger larvae and then possibly some eggs. Surrounding this brood should be an arc of stored pollen, and above this will be sealed honey – these are the food stores that the bees will draw on to nourish the nearby larvae. Usually at the bottom of the comb there will be larger cells, containing drone larvae. When capped, these cells will be domed.

The queen typically lays the first eggs of the year in a small patch which expands out across the comb over the following days. When there is no more room on the frame, she will move to those on either side of it. There will be a stage on each frame when the first eggs laid will have hatched, developed and pupated and new bees will have emerged. The queen then returns to the newly emptied cells and, after they have been prepared by workers, will lay in them again. There is therefore a constantly changing cycle of empty cells, eggs, larvae and sealed brood. Early in the season when the colony is expanding, there will be a higher proportion of eggs and larvae. Towards the end of the season when the colony is contracting, there will be more sealed brood. As you become more experienced you will be able to 'read' each frame and know how the colony is developing and whether there are any management decisions you need to make as a consequence.

As well as determining the development status of the brood nest, the other two important things to look for are signs of disease and indications of swarming, both of which are covered in the following pages.

Finishing the inspection

When you have been through all of the frames in a brood box, they should be replaced in their original position. Place the last frame that you inspected back against the far end of the box from where it was removed. The other frames can then be pushed back into place. If you are using Hoffman frames as recommended, this is easy; they can be slid back to their original position in blocks of three or four touching frames. Avoid separating frames when doing so, because this would create a gap for bees to fill, meaning they will have to be moved out of the way with the hive tool or a puff of smoke to avoid squashing them as the frames are pushed together. Work your way across the hive, pushing blocks of frames back into place until you are left with a gap at the far end. Re-insert the removed frame and finally the dummy board. Use your hive tool to lever frames closer together if you need to make a bit more space for the dummy board.

Closing the hive

Before closing up, do a little housework. If your bees have built brace comb along the top bars of the frames, run the blade of your hive tool along them and scrape it off. This will help to maintain bee space and prevent bees from getting squashed as the hive is reassembled. Removed wax should go in a sealed container kept for the purpose. Wax should not be left on the ground near hives as it can attract wax moths and robber bees. Any squashed bees and excess propolis around the edge of the box can also be scraped away.

Replace the queen excluder, using a puff of smoke if bees need to be cleared out of the way first, followed by any supers. The best way to replace any pieces of hive equipment is to carefully place them on the item below at an angle so that there are only small points of contact. The item can then be gently twisted into place, giving any nearby bees chance to get out of the way. Sadly, the occasional bee is likely to get squashed during inspections, but being careful and observant will help to avoid this. Replace each item in the order that it was removed, finishing with the hive roof.

When you first inspect bees you are bound to be a little nervous, inefficient and possibly a little clumsy. With a little experience, you will be able to move through a hive carefully and efficiently, with deft and dextrous movements. The bees will show their appreciation and the experience will be an enjoyable one for you too.

BELOW Scrape comb off the top bars before closing the hive.

Her Majesty

It is not necessary to see the queen every time you inspect your bees, although there is always pleasure in spotting her and knowing that she is well. Even without seeing the queen, you will know that she is in the hive if you see eggs, which must have been laid within the last three days. Occasionally, queens can go off lay for a few days, so no eggs doesn't necessarily mean no queen. If there is no queen, bees are usually more agitated or defensive. A queenless colony also emits a noticeably edgy-sounding roar rather than the usual contented hum of happy bees. There are other signs that a colony is queenless, which are discussed on page 138. A colony in which a queen is present, healthy and laying eggs is said to be 'queenright'.

Keeping track of the queen

There are a few situations when it is important to be able to find the queen. To help with this, queens are usually marked with a coloured dot of paint on the thorax. A different colour is used every year on a five-year cycle, so that whenever you see a queen you can tell her age. The dot is usually applied using a water-based paint pen, although some beekeepers use stick-on coloured or numbered discs.

All beekeepers follow the same colour convention, although there are various mnemonics to help remember the right order:

Year ending	Colour	Mnemonic
1 or 6	White	Will
2 or 7	Yellow	You
3 or 8	Red	Raise
4 or 9	Green	Gentle
5 or 0	Blue	Bees?

Finding queens

Having a marked queen makes her easier to spot, but this does mean that she has to be found, caught and marked in the first place – something many new beekeepers find daunting. If you buy a nucleus colony of bees when you begin, the queen should already be marked, but you are likely to find yourself having to

ABOVE A queen born in a year ending with a 2 or a 7.

mark a new queen yourself within a couple of years. Do not try to mark young queens until they are laying well. Before then they can be fast, flighty and hard to catch.

If you will be looking for the queen, use little or no smoke when opening a hive. Smoke tends to make queens more timid. Queens are usually to be found on frames where there is brood, which means you can separate the food frames from the brood frames and concentrate on looking only where she is most likely to be. Lift likely frames out of the hive as normal, keeping them over the hive in case the queen falls off or you drop the frame. Look first on the inner side of the frame that was not exposed to light – queens tend to move to where it is darker. Searching needs to be methodical. Scan the frame systematically, looking at the edges first as she is often spotted crawling around the edge to get to the darker underside. After checking the edges, gradually work inwards. Look at one side of a frame, check the other side and then repeat. If you don't see the queen, replace the frame and move on to the next one. If you don't find her after going through all the brood frames, give up and try again on your next visit – there's no hurry to mark a queen.

As you become more experienced you will find it easier to spot queens. It's a case of getting your eye in. Queens have a unique presence on a frame. They have a more purposeful, gliding motion compared with the other bees, their abdomens swish in a sinewy way, and they have longer, more spidery legs. Often you can spot a queen by the pattern of the workers around her, simultaneously moving out of her way to make space but also showing an interest in touching and licking her, sometimes resulting in a discernible halo arrangement of bees around her.

Marking queens

Experienced beekeepers catch queens by hand and hold them in between their fingers while marking them. This is a delicate operation and best done once you are more confident. Newer beekeepers usually catch and mark a queen using one of the commercially available queen-catching devices. The crown of thorns type can be lowered over the queen and pushed into the comb to trap her in place. The plunger type uses a sponge on a stick to guide the queen into a tube, allowing her to be lifted away from the comb once caught. Both devices are cheap, so buy one of each and see which you get on best with. It's worth practising catching and marking a few drones before you try this with your precious queen.

1. Catch the queen by trapping her under a crown of thorns or inside a plunger-type queen catcher.

2. Press the queen gently up against the mesh so that she is unable to move and her thorax is positioned between a gap in the mesh.

3. Shake the marking pen vigorously, then dab a few spots of paint onto your glove to get rid of any excess before gently pressing the nib onto the queen's thorax to mark her. Another way is to dab some paint onto your glove and then transfer a spot of it onto the queen using the end of a matchstick.

4. Once marked, allow the paint to dry for a minute before releasing the queen back into the hive.

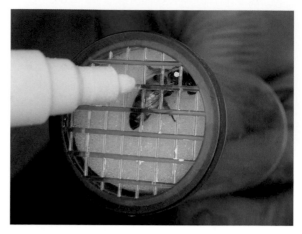

ABOVE Marking a queen in a plunger-type queen catcher.

Keeping records

It's important to keep accurate records of what you see and do each time you inspect a colony. This helps you to assess the health and development of the colony and make decisions about how to look after your bees, both in the near future and in years to come.

Most people use a standard record card for each colony. These can be purchased, downloaded or homemade. They usually comprise a list of checkpoints along with room for additional notes. The cards can be filed in a ring binder or kept in a plastic wallet pinned inside the roof of each hive. Some people prefer to write up each inspection in a notebook, like a diary entry.

Records can be completed as you inspect a colony, using the process as a reminder of each of the aspects that you want to look at, or you can write them up afterwards over a cup of tea – but if you have more than one colony it's surprisingly easy to get in a muddle. Colony record apps are available for smartphones or tablets, although using them at the hive when wearing sticky gloves is a recipe for disaster. If you are very organised and have the time, you can transfer your handwritten notes to a spreadsheet or other digital format.

Among the most important records to make are accurately dated observations that relate to the swarming process (see page 85). Other important factors are colony temperament, disease resistance and productivity. These are particularly important when it

ABOVE Good record-keeping is essential.

The following are some of the categories on a typical record card. New beekeepers might want to record all of these points and more. Experienced beekeepers tend to record much less information, often developing their own system of symbols or shorthand to make the process quicker.

1. Colony name or number
2. Date and time of inspection
3. Weather conditions and temperature
4. Temperament of colony
5. Queen seen?
6. Play cups or queen cells seen?
7. Any signs of pests or disease?
8. Number of frames of brood
9. Number of frames of food
10. Room for colony to expand?
11. Number of supers on hive
12. Actions taken
13. Actions and equipment needed before next visit

comes to rearing new queens – your records will show which are the best colonies to raise new queens from, allowing you to improve the quality of your stocks over time.

To help with record-keeping, it is useful to name or number each colony, but remember that many qualities of a colony relate directly to the queen. Therefore, if the queen moves to a different hive, the records should go with her. For example, if you catch and re-hive a prime swarm from one of your colonies, the records for that colony should be transferred to the new hive because that is the new home of a previously recorded queen. The colony from which the swarm escaped should soon contain a new queen and, although she is the daughter of the original queen, the characteristics of her colony may differ, so a new record should be started.

Most categories on a colony record card can be completed using a tick, a number or a score. For example, a tick might confirm that the queen has been seen, a number might relate to how many frames of brood are in the colony, and a score out of five might indicate how well behaved the bees were during your inspection.

As well as recording what you see in each colony, it can be worth recording additional incidental information. For example, noting the time when various plants and crops flower each year will give you a sense of whether a season is a little early or late, helping you to forecast when swarming is likely to begin, or a honey flow might occur.

Because honey bees are considered to be a food-producing animal, in the UK it is now a legal requirement to keep records for five years following the use of any form of medication. You must record the colony name or number, the name of the treatment, date of use, batch number and method of disposal, among other details. Visit the National Bee Unit website to download a medicine record card with all necessary information.

How to manage swarming

Swarming

The comic book image of swarms as boiling clouds of bad-tempered bees chasing terrified, arm-flailing victims is familiar but far from accurate. Swarms are not angry, threatening or dangerous. They will not chase anyone and you would be very unfortunate to get stung. To the beekeeper, swarms can be a thrilling and awe-inspiring sight. They are, however, inconvenient, unproductive and bad for public relations. Fortunately, they are largely preventable, and in the summer months swarm management is among the beekeeper's foremost concerns.

Swarming is the result of the natural desire of a healthy colony to sustain its genetic line. Honey bees reproduce on two levels. The queen lays eggs and creates thousands of young bees. In this way, the colony grows and prospers. But when a healthy colony is expanding it might choose to improve its chances of survival by dividing into two or more parts, with some of the bees leaving to set up home elsewhere. Rather than individual reproduction, this is reproduction at the level of the colony.

Swarming is a desirable and positive outcome for a colony of honey bees – but not for the beekeeper. When a colony swarms, you will lose about half of the bees, along with your established, laying queen. This will lessen the likelihood of a honey harvest, because there might be too few bees left to collect and store a surplus. Furthermore, if left unchecked the remaining bees could swarm again within a few days, or they might fail to raise a viable new queen, placing the very survival of the colony at risk. However, the greatest damage might be to your reputation – however sympathetic they may be to your hobby, few neighbours will welcome menacing-looking clouds of bees swirling over the fence.

OPPOSITE PAGE Swarming bees, one of nature's wonders. Don't run away, but stand and marvel.

Colonies therefore should be inspected frequently during the summer. If signs of swarm preparations are seen, there is limited time in which to manage the situation and achieve a result that is satisfactory for bees, beekeeper and neighbours.

How bees swarm

To divide a colony into two units, each with a chance of survival, worker bees need to ensure the availability of at least two viable queens – one to leave with the swarm and one to remain in the nest. For a colony to swarm, therefore, new queens must be produced. Workers prepare special cells called queen cups to receive eggs laid by the existing queen. The resulting female larvae are fed large amounts of royal jelly and develop into queen larvae, growing rapidly until they start to pupate on the eighth day after their egg was laid. Worker bees then cap their cells, sealing them until the new queens emerge another eight days later. Thus the time from an egg being laid until the emergence of a new queen is 16 days. These timings are important to remember when trying to manage your bees' attempts to swarm, so there will be several reminders of them later.

As soon as a queen cell is sealed and a queen larva begins to pupate, the colony 'knows' it is in a position to swarm. It has a mature, mated queen with whom some workers can establish a new nest elsewhere, and there are one or more new queens in an advanced state of development that will soon emerge to head the existing nest. Worker bees who will soon leave the nest in a swarm begin eating honey, filling themselves with an important, portable resource. On the day of swarming, worker bees exit the hive and fly around the entrance, many of them collecting on the front of the hive. The old queen is herded out of the hive – this is an operation controlled by the workers – and takes flight along with the rest of the swarm. About half of the bees in the colony are now on the move, in a huge, swirling, thrillingly noisy cloud. The swarm moves through the air until it reaches a nearby object, usually a tree or bush typically no more than about 20 metres from the hive. The bees settle on the object and quickly coalesce into a dense cluster about the size and shape of a rugby ball, although a swarm from a large colony can be much bigger. There is a curious stillness to a clustered swarm. What minutes earlier was a noisy, chaotically purposeful throng becomes a sedate, barely whispering ball of bees that hangs like some strange fruit. This is when a swarm can be caught and re-hived (see page 63). However, a settled swarm is surprisingly easy to miss, even when you are standing very nearby.

Finding a new home

If you watch a clustered swarm – and it is usually perfectly safe for you to inspect them at close quarters – you will see some worker bees scurrying around on the surface and dancing. These are scouts, communicating information about possible new nest sites that they have found. While the bees were swarming and clustering, and possibly for some days beforehand, these scouts will have been flying for several miles in all directions, looking for somewhere new to live. Scouts visit hollow trees, chimneys, post boxes, sheds and roof spaces, and possibly even empty beehives, considering factors such as cavity volume, size and orientation of the entrance hole, height from the ground and so on. Having assessed the quality of a potential home, scouts return to the swarm cluster and use the language of dance to communicate what they have found (see page 114). The dance not only tells other bees where the vacant property is, but also expresses an opinion about its suitability. Having received directions, other bees will visit the site and, if in agreement, will begin to dance in its favour too. Swarm clusters can hang on a branch for an hour or two, or sometimes several days, but when enough bees dance in favour of one location, a decision is reached. Suddenly, the bees take to the sky and the swarm flies directly to the chosen site. Thousands of bees and their queen flood through the door, settling rapidly into their new home, where they immediately begin the process of building new comb, gathering stores and raising brood.

Emergence of new queens

Anything from one to more than a dozen capped queen cells will have been left developing in the nest from which the swarm came. There is a degree of unpredictability about what happens next. If the colony still contains plenty of bees, it might swarm again with the first new queen, once she emerges from her cell about a week after the first swarm left. In such cases, the virgin queen swarms with around half of the remaining bees – perhaps a quarter of the original colony. These smaller swarms are called casts, or after-swarms. The first swarm to have left with the original queen is called the prime swarm. Casts are less likely to survive than prime swarms because they comprise a reduced workforce and are headed by a virgin queen (or sometimes several) who has yet to mature and successfully mate – in itself a risky proposition. Following the departure of a cast, the next queen to emerge might also swarm, this time with even fewer bees than her predecessor. Cast swarming can continue like this until tiny swarms comprising just a handful of bees and a queen leave the colony. In such cases, neither the cast swarms nor the heavily depleted colony remaining in the original nest are likely to survive.

Alternatively, the first queen to emerge, or perhaps the second, will stay in the nest and lay the eggs necessary to build the colony back to full strength. To do so, she must get rid of any rivals. This is achieved by emitting a squeaking song known as piping. Innocently, other queens still in their cells reply, betraying their presence and location. The first queen then finds them and stings them to death. This is the only time when a queen is likely to use her sting, which – unlike the barbed sting of a worker bee – is smooth and reusable.

Swarm prevention

Not every colony will swarm every year, but there are several factors that make swarming more likely. By being aware of these, beekeepers can manage their colonies to reduce the likelihood of swarming. The following factors affect the likelihood of swarming:

Space

This is one of the prime factors. A colony of bees without much space to expand will, naturally, think about moving home. When a colony is expanding rapidly in spring, a

ABOVE Queen cells are often built on the edges or corners of frames.

ABOVE Queen cells can easily be hidden beneath bees.

ABOVE A crushed queen cell showing developing pupa.

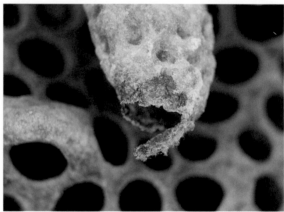

ABOVE The trap door on an emerged queen cell.

hive can quickly become congested. In particular, with the queen laying 1,000 or more eggs a day, empty cells soon become scarce. This is exacerbated by increasing amounts of nectar being brought into the hive as more plants come into flower. The nectar, and honey produced from it, have to be stored somewhere, thus filling cells in the brood nest that could otherwise be used by the queen to lay eggs. One indicator that more room is needed is brace comb being built along the top bars of the frames in the brood nest, showing that the bees are trying to create more storage space.

Congestion can be minimised by increasing available space. This is done by adding supers with frames where nectar can be stored. It can be difficult to know when best to add a super, but if around mid-spring the bees in a brood box are occupying more than half the frames,

it's probably time to add a queen excluder and a super. When you first start keeping bees you will have to add a super full of frames containing foundation. The bees will draw this out to produce comb – a demanding activity that is thought to help delay swarming. After a few years you will have super frames already containing built comb, which the bees will be able to use straight away. If using these, it is worth adding a few frames of foundation to keep the bees busy making new comb.

Another way to give bees more space is to divide a colony before it begins preparations to swarm. Colonies that become very populous early in the season can be split in half, with one half having the queen and the other half being given sufficient eggs or young larvae to raise a new queen of their own (similar to the nucleus method of swarm management, see page 87). Such pre-emptive measures mean that you might not get a spring honey crop, but the colonies could expand sufficiently to take advantage of the summer honey flow.

Age of the queen

Generally, when you start beekeeping and have a new, young queen, swarming is not something to worry about in your first season. In your second season, the queen will be older and the colony is more likely to swarm, which means that you must be more vigilant. If there are no attempts to swarm in a queen's second year, be prepared for swarming the following season.

Swarmy bees

Some strains of bees are swarmier than others and seem to want to go at the earliest opportunity. When you have several colonies and can choose which ones you will raise new queens from, propensity to swarm is one of the main qualities to consider. Commercially available hybrid bees often have a low propensity for swarming.

The signs of swarming

Whatever measures you take to discourage them, most healthy colonies will try to swarm at some point. Regular colony inspections will allow you to see the signs of swarming and take the appropriate action.

Colonies will not swarm until there are sufficient drones for young queens to mate with. So, until you see plenty of drones in your colonies, you needn't worry about swarming.

The first visible sign that bees might be preparing to swarm is the appearance on brood frames of small, acorn cup-like cells called queen cups. Built vertically on the comb with their opening at the bottom, these are the foundations of queen cells in which the queen might lay an egg destined to become a new queen. Often queen cups are built and then torn down a few days later, or they are made and left unused. They tend to be built around the edges and bottom of combs. If you see them, make a note that the colony might want to swarm.

ABOVE A very populous colony – swarming is likely soon.

The next visible sign is preparation of the queen cups to receive eggs. Hold the frame upside down and look into the cups, or break away the sides with your hive tool. If the inside base is dark, smooth and shiny, the workers are polishing it with propolis to encourage the queen to lay an egg.

The presence of an egg is the next clue but, frustratingly, is not conclusive. If a queen lays in a queen cup it is likely that the subsequent larva will develop into a queen, but sometimes the workers remove the eggs – perhaps because conditions are not currently suitable for swarming. However, if you see eggs in queen cups, it is safest to assume that your bees are preparing to swarm. If you find larvae in queen cups, indicated by the presence of glossy, creamy-white royal jelly, the bees are undoubtedly preparing to swarm. At this stage you might also notice that the outer walls of the cup are beginning to be extended downwards to produce a queen cell.

ABOVE A drone emerging. Swarming is possible.

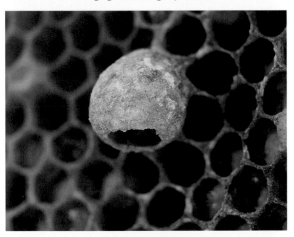

ABOVE Queen cups are an early sign of possible swarm preparations.

Regular inspections and critical timing

To spot the signs of swarm preparations and prevent unwanted swarming, you should inspect the brood nest of your colonies regularly during swarming season. Exactly when bees might swarm can vary yearly according to climate and location, but it typically occurs at any time between late spring and mid-summer. The safest policy is to start inspecting colonies regularly once you see the appearance of drones. Drones are not produced in the earliest parts of the season, only appearing in time to be available to mate with queens. Once you see drones in your colonies, you should aim to inspect them every seven days. This schedule is based on the development times of new queens – information already given, but worth repeating.

When a queen lays a fertilised egg, it will hatch three days later to produce a female larva. All female larvae are the same, whether born in a normal brood cell or in a specially prepared queen cell. It is the quality and quantity of food given to those in queen cells that induces them to develop into queens. As queen larvae rapidly grow, workers extend the walls of their cells from the initial, shallow, acorn-cup form into extended tubular structures about 3cm long. When

ABOVE Larvae in queen cups means that swarming preparations are well under way.

85

queen larvae are just five days old, eight days after their egg was laid, they begin to pupate. At this stage worker bees seal the queen cell, shutting the pupating queen safely away. There is now at least one viable, developing queen ready to emerge into the nest in eight days' time – 16 days after the egg was originally laid. Usually, bees waste no time in swarming soon after the first queen cell in a nest has been sealed. Swarming can therefore happen eight days after an egg is laid in a queen cell, and five days after that egg has hatched. It is an astonishingly rapid process.

If you find sealed queen cells when you inspect your colony, the chances are that your bees have already swarmed. Therefore, you should inspect your bees before this can happen. Since it takes eight days from an egg being laid to a queen cell being sealed, you should leave no more than eight days between inspections. Fortunately, this works perfectly with our seven-day week, and makes it easy to remember that you need to inspect your bees on the same day each week – every Saturday, for example. If you can't do it on Saturday, Friday is a good alternative. Leaving it until Sunday is taking a risk.

Clipping queens

If seven-day inspections will be difficult to manage all the time, or you are worried about missing the clues and losing a swarm, it is possible to clip the queen's wings so that she cannot fly. If a colony tries to swarm but the queen does not fly with them, they will return to the hive. A clipped queen might fall on the ground and be lost when trying to leave, or might manage to return to the hive, but the majority of your bees will remain. However, the colony is still expecting to swarm and will leave instead with the next available queen – the first to emerge from her queen cell, typically eight days after the originally planned swarm date. This means that even if your bees attempt to swarm, with a clipped queen you could leave an inspection for two weeks and expect to find all your bees still present – except perhaps the queen. Clipping queens does not stop swarms but does buy you some time. When you purchase your bees, the vendor should be able to clip your queen's wings at your request. Clipping seems to have no ill effects on the queen, although it does rather spoil her beauty. It is a matter of personal choice.

ABOVE Clipping a queen can lessen the chances of losing a swarm but does diminish her regal splendour.

Swarm management

Inspecting your bees every seven days means that you should spot the signs of swarm preparations and be able to manage the situation, preventing the loss of a swarm and your queen. The main signs are queen cups with larvae in them, or growing but not yet sealed queen cells. The key to successful management of the situation is having a pre-determined plan. The plan should include a list of actions and the appropriate timings, as well as ensuring you have sufficient spare equipment to do the job. Whichever method of swarm management is used, the same basic principles come into play. If these principles are remembered, any number of techniques can be successful.

Think of a colony trying to swarm as comprising three parts:

- The queen
- The brood (eggs, larvae, sealed brood, queen cells and young, non-flying bees)
- The flying, older bees

Most methods of swarm management rely on separating any one of these elements from the other two, to make swarming unfeasible. For example, if a queen is removed from a colony, it cannot swarm because the bees cannot leave and set up home elsewhere without her. Likewise, if the brood is removed from a colony, it will not swarm because it will not be leaving behind sufficient resources (eggs and larvae) to ensure the survival of the colony.

Another key principle that makes it possible to manage colonies preparing to swarm is the way that flying bees will always return to the place that they remember as home. If a hive is moved several metres or more to one side, the flying bees will exit and return to the space where they think home should be (see page 99). This trait provides a useful way to separate flying bees from a colony when trying to prevent them from swarming, as used in the method described below.

Swarm management methods

Swarm management methods can be baffling to new beekeepers – and to some experienced ones as well. There is the Snelgrove method, the Demaree method, the Pagden method, and others too. All have their advantages and disadvantages, and their enthusiasts and detractors. It really doesn't matter what method you use as long as you understand what you are doing and achieve the result that you want. Beginners' courses usually teach the so-called Pagden artificial swarm method, because it is often thought to be a good way for new beekeepers to fully understand the nuts and bolts of swarm management. In my experience, though, this method often confuses beginners and leaves them feeling intimidated by the complexity of it. For new beekeepers, the nucleus method is the best option. It's easy to understand and perform, and it requires the least additional equipment. It is probably the method most used by beekeepers, but once you have got the hang of it, you can try something different and see which method you prefer.

The nucleus method of swarm management

This method involves removing the queen from the colony that wants to swarm and placing her in a nucleus box, thus separating her from the brood and the flying bees. Because nuc boxes are small and light,

ABOVE Swarms are beautiful; take the time to study them closely.

it is relatively easy to move the separated queen and her attendant bees to another part of the garden or, preferably, to a different apiary. If you don't have a nuc box, the procedure can be carried out with a spare full-sized hive instead. At the end of the procedure, you should have two colonies; a nuc containing young bees and brood along with the old queen who will continue laying eggs and building up a new colony, and the original hive containing a slightly smaller colony with the older bees and with a replacement queen in the making.

WHAT YOU WILL NEED

- An empty nuc box or hive
- Enough frames fitted with foundation to fill the nuc box
- Drawing pins
- A spare hive stand

TIP: If during inspections you find unsealed queen cells in a colony, don't panic. Close the hive, have a cup of tea, go over your plan and then gather your equipment. When you are sure you know what you are going to do, proceed with the operation.

1. Place your nuc box next to the colony that is preparing to swarm (the parent colony). Stuff the entrance of the nuc box with green, sappy grass. Take out the frames of foundation.

2. Remove a frame of stores (honey) from the brood box of the parent colony. These are usually the outer frames, and removing one makes room to move the other brood frames around. Check carefully whether the queen is on the frame of stores (unlikely), and then place it, along with any adhering bees, against one wall in the nuc box.

BELOW Knocking down a queen cell.

3. Go through the parent colony, carefully and slowly looking for the queen. If there are no sealed queen cells, the colony should not have swarmed and the queen should still be present. Try not to use much smoke as this can make queens hide. When you find the queen, she will probably be on a frame of brood. Check very carefully that there are no queen cells on the frame with the queen. Pay particular attention to the edges and bottom of the frame. If you find queen cells at any stage of development – or anything even like them – squash them with your hive tool. Place this frame into the nuc box, carefully pushing it up against the frame of stores already there. Make sure that the queen remains on the frame and goes safely into the nuc box.

4. Next, find another frame of brood in the parent colony, ideally one with plenty of sealed brood, some of which has young bees emerging from it. This will ensure that the nucleus colony will soon have lots of young nurse bees to care for the next generation of brood. The emerging brood will also leave empty cells for the queen to lay more eggs in. Again, check very carefully that there are no queen cells on the frame and place it into the nuc box, pushing it gently up against the frame with the queen.

5. Return to the parent colony and find one brood frame that has a healthy-looking open queen cell. (If you spot one before you find the queen, it's worth taking note of it and coming back to it.) This cell will develop into your future new queen. Don't be tempted to choose a sealed cell in the belief that you will have a new queen a few days earlier – you will almost certainly regret it. Choose a queen cell in which you can see a fat, glistening larva in a nice pool of royal jelly. A cell that is tucked into the side or corner of a frame is ideal because it is less likely to be accidentally damaged when you return it to the hive. Before replacing the frame, thoroughly search it for any other potential queen cells and destroy them with your hive tool. Be very careful when handling this frame, because you don't want to dislodge your chosen royal larva from its bed of jelly. Place the frame carefully back into the parent colony and push a drawing pin into the top bar above the chosen cell so that you can easily locate it in the future.

6. Being careful not to disturb or damage your chosen open queen cell, remove two more brood frames from the parent colony. One at a time, lower them into the gap in the nuc box and give the frames a hard, sharp shake so that most of the bees fall into the box. Cover the nuc box in between the shaking of the frames to ensure that most of the bees stay inside. By doing this you are giving the nucleus colony a good number of young bees to help raise the next generation of brood. You shake bees from brood frames because this is where most young bees are found. Having not flown yet, these young bees are more likely to stay in the nuc with the queen. After shaking bees from the frames, look for anything that resembles a queen cell and squash it before returning the frames to their original position in the parent colony.

7. Now find one more frame of food and place it with adhering bees in the nuc box next to the second brood frame. The order of frames in the nuc box is now one frame of food, two frames of brood (plus queen), a second frame of food. Place frames of foundation in the box to fill the remaining gap. Close up the nuc box.

8. Next, go through any remaining brood combs in the parent colony and remove anything that looks like a queen cell. Shaking the bees off each frame will help to reveal any. Sometimes open drone cells can look a little like queen cells in the making. If you aren't absolutely sure, squash these as well. Rearrange the frames in the parent colony so that the brood frames are in their original order. Flank these with frames of foundation removed from the nuc box. Then replace any combs of stores and close up the hive.

9. Now turn your attention to the nuc box. You can put it on a stand near to the parent colony or elsewhere in the garden. When the green grass blocking the entrance wilts, most flying bees that were transferred to the nuc will leave and return to the parent hive, leaving mostly nurse bees in the nuc. If you can take the nuc to another apiary more than 3 miles (5km) away, all of the flying bees will reorientate to the new location and more of them will be retained. With a laying queen, plenty of young bees, brood in various stages and food, this colony should build up quite rapidly and may need to be moved into a full-sized hive in a month or so. Keep an eye on it in case the bees run low on stores and have to be fed.

ABOVE Nucleus method swarm management in progress.

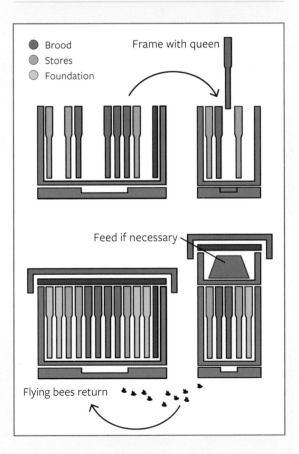

What happens next

In the parent colony, the bees will quickly sense that their queen has gone and will begin to make more queen cells, using any larvae up to three days old. These will be emergency queen cells (see page 93). Any larva not initially destined to become a queen still has the potential to become one until its third day of life, because until then queen and worker larvae receive the same rich diet. From its third day, a worker larva is fed a different diet and the opportunity to turn it into a queen is lost. If a three-day-old larva is present when you undertake the operation, it could be promoted to a queen larva and become a sealed queen cell within just five days.

If there are eggs in the comb on the day of the operation, they too could be chosen to become queens for up to the next six days (the three days it takes them to hatch, plus the first three days of their larval life). This means that, after six days, the worker bees cannot choose to make any additional queens because there will no longer be any larvae young enough to convert.

So, seven days after splitting your colony it is vital to open the parent colony and inspect each frame, removing every newly made emergency queen cell. Emergency queen cells are small and can be tricky to spot underneath bees, so shake the bees off every frame and remove any sealed or unsealed queen cells that you find. You will need to do the same with the frame marked with a drawing pin and containing your selected queen cell, but brush rather than shake the bees off so that you don't damage your soon-to-emerge queen. Her cell will now be sealed, but she could still be harmed by rough handling.

After this you can relax, safe in the knowledge that you have removed every queen cell in the colony, leaving only your chosen, sealed cell. No more queen cells can be made by the bees because there are no eggs or larvae of suitable age. A few days after your inspection, the new queen should emerge into the colony from your chosen queen cell. She should spend a few days maturing before going out on her mating flight, then returning to start laying eggs and building the colony back to full strength.

Leave the hive alone for three weeks after leaving the single, sealed cell in the colony before checking to see if your new queen is laying eggs. Once she is, you can allow the colony to build up on its own or choose to unite it with the bees and brood that were placed in the nuc. If your new queen isn't laying within a month, test for queenlessness with a frame of eggs and brood taken from another colony (see page 138). If no new queen is present, you will be glad you still have the original mother of the colony in a nucleus and ready to be reinstated.

Sealed queen cells

If during inspections you find sealed queen cells, you might already have lost a swarm. First check whether the queen cells that you see are indeed swarm cells. They could instead be emergency or supersedure cells (see pages 93–94). If you are certain that what you have found are sealed swarm cells it is likely that your colony has already swarmed – although this is not guaranteed. You would think that with perhaps half of the bees having left it would be obvious that there were fewer in the hive, but to the inexperienced it isn't always easy to tell. The most reliable indicator is the presence or absence of the old queen. If you are lucky, she will still be there, and you have caught things just in time – they might have gone within a few hours. If you do find the queen, place her in a nuc and proceed with the nucleus method of swarm control as described earlier, making sure you remove any sealed queen cells from the colony and that you leave a single open one.

If the queen is not present, inspect nearby trees and other possible landing sites – your swarmed bees might be hanging nearby ready for you to retrieve them. After dealing with any swarm (see page 63), next turn your attention to the parent colony. If the queen has gone and you find sealed and unsealed queen cells, remove all of the sealed cells, leaving behind a single unsealed one. Return in seven days (before any new queens can emerge) and remove all but one sealed queen cell. No more queen cells can be made, and the one that you leave will produce a new queen to head the colony.

If you find only sealed queen cells in your colony, and there are no eggs and only quite large larvae in ordinary worker cells, it means that the swarm departed several days ago. Remove all but one large, well-formed queen cell. The remaining worker larvae are probably too old to be made into emergency queens, so only the queen cell left by you can emerge and head the colony.

ABOVE A large prime swarm.

ABOVE A small swarm, probably a cast.

Check carefully that any sealed queen cells that you find have not already emerged, and that your colony is not already sending out cast swarms. Queen cells from which a queen has emerged will be open-ended and empty, or sometimes have a small open trap door hanging from the bottom, through which a queen has emerged. This means that there are likely to be one or more virgin queens already in the hive. If you find emerged queen cells, inspect all of the remaining sealed cells and try opening any ripe ones to release the fully developed queens. Ripe cells are ones where workers have nibbled

away the wax from the tips to reveal the brown papery cocoon inside, making it easier for the queen to emerge. Use a sharp hive tool or knife to carefully cut open the papery cocoon. Virgin queens will energetically squeeze their way out and quickly scurry onto the comb and mix with the workers. Release as many virgins as you can, and then remove all other remaining queen cells. As a result, the released queens will usually fight and one will survive, going on to head the colony. Two weeks later, check to see if the new queen has mated and is laying eggs, although this can sometimes take as long as five weeks.

Queen and apiary management

Queen management

Queen cell variations

Queen cells at various stages are usually taken as a sign that a colony is going to swarm, or possibly already has. In fact, there are three reasons why a colony might construct queen cells, swarming being only one of them. It is important to be able to identify the cause of any queen cells that you find, and what they can tell you about the condition of the colony.

Swarm queen cells

These are constructed when a colony wants to reproduce by swarming. They start as specially built queen cups that protrude vertically from the comb with a downward-pointing opening. As queen larvae develop, worker bees extend the cell walls and sculpt them, producing an appearance reminiscent of a peanut shell about 3cm long. Swarm queen cells usually hang from the side or bottom edges of a comb, or are attached to the bottom half of the face of a comb. Typically, somewhere between six and 20 cells are produced.

Emergency queen cells

These are built when a colony suddenly and unexpectedly finds that it has no queen. This could be because the old queen has died naturally, or perhaps was accidentally squashed by the beekeeper during the last inspection.

Within minutes of noticing that they are queenless, workers take emergency measures to produce a new queen. This is known as the emergency impulse. If there

LEFT A queen in her prime.

ABOVE One capped swarm cell, and two nearly complete ones.

ABOVE Stubby-looking emergency queen cells.

are eggs in the hive, some of the resulting larvae will be raised as queens. Existing larvae under about three days of age can also become queens if they are continuously fed the rich diet that stimulates the development of a queen rather than a worker.

Emergency queen cells look different from swarm queen cells because they were not originally intended to house queens. The eggs or larvae do not start life in specially constructed vertical queen cups, but instead lie at the bottom of normal horizontal worker cells. As the promoted queen larvae grow, the workers extend the walls of their cells horizontally outward and then curve them downwards on the surface of the comb. Emergency queens therefore develop either at an angle or possibly slightly curved inside the cell. The visible portions of the cells are usually about half the length of swarm queen cells and are less well sculpted. Typically, about a dozen are made, wherever on the comb eggs or young larvae are available, frequently in the top half of the comb and on several frames.

If you find emergency queen cells, it means you have lost your queen. All you can do is leave the cells and let

ABOVE Discreetly placed supersedure cell.

the bees raise new queens, one of which will mature and head the colony. Although there may be many cells, the colony will not swarm.

Supersedure queen cells

A good-quality queen is the prime concern of workers in a colony. If they find a queen unsatisfactory, perhaps because she is injured, lays too few eggs or is getting old and might not survive the winter, they can create a replacement queen. In such cases a small number of queen cells are made, and one of the resulting young queens is protected by the workers until she has matured, mated and started to lay. Usually, only when the new queen is established is she allowed to come into contact with her mother, whereupon they will fight, and the older queen is likely to be killed. Sometimes mother and daughter will co-exist for a while, but at some stage the daughter takes over. Many beekeepers value supersedure because it can mean the seamless replacement of an old queen by a young, robust one of the same strain.

Supersedure queen cells tend to be built on the face of the comb and, although large, can be less prominent than swarm queen cells. They are often towards the top of the comb, perhaps because the old queen is less likely to find and destroy them there – although workers do seem to protect such cells from possible attack. If supersedure is to replace an old queen, it tends to be done between mid-summer and late summer so that the new young queen can take the colony through winter.

If you find suspected supersedure cells, they are best left so that the colony can get on with raising a replacement queen as they wish. You do need to make sure, however, that they are not swarm cells, which is somewhat easier said than done.

Re-queening

Everything about the character of a colony comes from its queen. Although queens come and go through natural processes such as swarming and supersedure, there are times when a queen might need to be replaced directly by the beekeeper. Reasons for this can be:

ABOVE Colonies in an apiary can vary in quality as a result of the age and attributes of their queen, which should be replaced when necessary.

- To replace an old or drone-laying queen (see page 138)
- To keep young queens and discourage swarming
- To remove a queen whose genes make a colony unpleasantly defensive
- To improve the quality of your bees generally, making them less prone to disease or better at producing honey, among other considerations.

There are various ways to obtain new queens. You can raise your own through the manual selection and transfer of larvae under carefully controlled conditions (queen-rearing), you can give a colony the opportunity and material to raise their own new queen using the emergency impulse (discussed later), or you can buy a mated queen from a trusted supplier.

Queen-rearing involves specialist skills and equipment, and can produce new queens in large numbers. It is fascinating and rewarding but is best done after several years of beekeeping experience. If you require just one or two new queens it is better either to acquire them from an outside source or, preferably, to give your bees the opportunity to raise their own.

Colony split method of queen raising

This method allows bees to raise their own new queen as a result of the emergency impulse when they find themselves suddenly queenless. It relies on having at least one colony which exhibits the qualities that you want in your bees. These qualities come largely from the queen, so daughter queens should have similar traits. The technique can be used from late spring to mid-summer, when drones are available for mating. The method described here involves replacing the queen in a colony whose qualities are not desirable, using eggs and larvae supplied by a colony with characteristics you want to replicate. It relies on you having at least two colonies. The same method can also be used with material from just one colony from which you wish to raise a new queen – in other words, if you currently have only one colony, you can use it to raise a second one. Once you are familiar with the concept behind these techniques you can begin experimenting with all sorts of variables to suit your own situation.

WHAT YOU WILL NEED

- An empty nuc box or hive and stand
- A queen cage
- Enough frames fitted with foundation to fill the nuc box
- Drawing pins
- Bee brush
- Feeder and syrup

1. Stuff the entrance of an empty nuc box with fresh, green grass so that no bees can escape. Over the next day or so the grass will wilt, releasing the bees.

2. From the colony to be re-queened, remove two good frames of stores containing plenty of both pollen and honey, as well as lots of bees. Make absolutely sure that the queen is not on these frames. It is safest to first find and cage the queen, putting her aside

3. Place the selected frames in the nuc box, leaving a gap between them.

4. Take one brood frame from the colony to be re-queened and place it in the nuc box, next to one of the frames of brood. This frame should contain mostly sealed brood. Again, make sure that the queen is not on the frame. Young bees will soon emerge from the capped cells, helping to populate the new colony that you are making, and to nurse the next generation of bees. The empty cells that they leave behind can be used by your new queen to lay her first eggs.

5. Take two more brood frames from the colony to be re-queened. Make sure the queen is not on them, and then shake the bees from these frames into the nuc box. Brood frames contain mostly young bees, which will remain in the new colony that you are creating without flying back to the old hive. Return these frames to the hive from which they came.

6. If you have caged the queen of the undesirable colony, release her back into the nest and close up the hive.

7. Select one brood frame from the other colony, whose desirable queen you want to mother a new queen. The frame should have plenty of eggs and very young larvae from which a new queen can be made.

8. Shake or brush every bee from the brood frame back into their hive – in particular, make sure that you are not transferring the queen. Then place the frame in the nuc next to the brood frame that is already there.

9. Push a drawing pin into the top of this brood frame to remind you which one it is.

10. Push the frames together and add frames of foundation (or drawn comb if you have it) to fill any gaps either side of the brood and food frames.

11. Ideally, take the nuc to a site at least 3 miles (5km) away. This means that the older, flying bees will not return home. Provided you have added plenty of young bees to the nuc, it can remain close to the parent hives – young bees will not fly back to their old home, maintaining a good population.

12. Check the two brood frames after seven days. The frame donated from the desirable colony, marked with a drawing pin, should have emergency queen cells on it. You do not need to destroy any of the queen cells on this frame. If you cannot find any queen cells, add another frame of eggs and young larvae, without bees, from the donor colony. The second brood frame, taken from the undesirable colony, should have some empty cells from which new worker bees have emerged. Carefully check this frame and remove any emergency queen cells that might have been built from eggs or larvae of the correct age that it contained when transferred.

13. Add a feeder and about a litre of sugar syrup if stores are running low, perhaps if the weather has been cold and wet. Keep an eye on food levels.

14. After finding emergency queen cells, close the hive and leave it alone for at least two weeks. Then check once a week until you find that a new queen is laying eggs.

15. When brood laid by the new queen is sealed and you are happy with her performance, use the newspaper method (pages 103–104) to unite this colony with the one whose queen needs replacing.

This method assumes that you do not want to deplete the colony containing the good queen. By removing only one frame of brood, the colony will remain strong and might still produce a crop of honey. If the desirable colony is already very large, or you fear it might swarm (but have not yet seen the signs of swarming) you could split the colony in half, with half of the bees, brood and stores remaining in the hive, the other half going into a nuc box. This avoids having to find the queen, because she must be in one half or the other. The half without the queen will produce a new one using emergency queen cells. This will give you two good-sized colonies that should grow rapidly, and it will reduce the likelihood of swarming. Should a new queen fail to be produced using either method, the two halves can be united (pages 103–104).

Another alternative is to split the undesirable colony in half, killing the unwanted queen when you find her. Give each half a frame of eggs and young brood from your desirable colony, and destroy any emergency queen cells that are made on frames with eggs or brood produced by the old queen. This gives you the chance to produce two good colonies from one undesirable one.

The disadvantage of these methods is that, although you are using eggs or larvae produced by your best queen, the resultant daughters must still leave the hive and mate with drones of unknown origin. Therefore, there is still an element of uncertainty about the quality of the bees that you will end up with.

Queen introduction

Rather than raising your own queens, you can buy mated queens from a supplier whose bees are of a known quality. They may have ensured their queens are mated in an area where they have been able to control the local drone population, or they might even instrumentally inseminate their queens using sperm from carefully selected drone stock. Try to buy from a local beekeeper who offers local-type bees rather than hybridised 'Buckfast' types, or imported queens. Buying local bees reduces the likelihood of subsequent generations being unsuitable and unpredictable, which perpetuates the need for regular queen replacement.

There are many ways to introduce a new queen to a colony, and there are even whole books dedicated

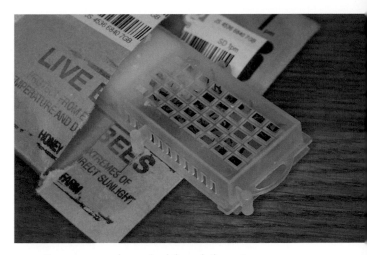

ABOVE New queens can be received through the post.

to the subject. Unfortunately, no method is 100 per cent reliable, and there is always a chance that your expensive new queen might be killed. One problem is that the pheromones produced by a new queen will be unfamiliar to the workers in the colony that she is introduced to, leading to rejection. This can be solved by disguising the chemical signals when the queen is first introduced, or by letting the bees become gradually accustomed to them. Another challenge is that bees are more likely to accept a new queen that is in the same physical condition as the one which has been removed. New queens, particularly those sent by post, can take a while to begin laying eggs, leading to rejection if they are used to replace a queen in full lay. The following is just one of many methods of queen introduction. It is slower than some, but has a high success rate.

Order a new queen from your supplier. She will be supplied in a small plastic travelling cage, along with several attendant worker bees who will care for her and feed her using sugar fondant that comes in the cage. Place one or two drops of water on top of the cage so that the bees can drink, and place it somewhere warm and dark, such as in an airing cupboard. Try to arrange the timings of the following so that the queen stays in her cage for as short a time as possible – one or two days is fine. You will also need a queen introduction cage, which is a square plastic or metal cage about the size of a slice of bread. An entrance to the cage is filled with fondant which must be eaten through before the queen can be released.

WHAT YOU WILL NEED

- An empty nuc box or hive and stand
- A queen introduction cage
- Fondant or marshmallow
- Enough frames fitted with foundation or drawn comb to fill the nuc box
- Drawing pins
- Bee brush
- Feeder and syrup

1. Stuff the entrance of a nuc box with fresh, green grass so that no bees can escape.

2. From the colony to be re-queened, select three frames of brood. One should have plenty of capped worker brood, while the others can have brood at all stages, including newly laid eggs. Take precautions to avoid transferring the old queen at this and subsequent stages.

3. Add the brood frames, including all adhering bees, to the centre of the nuc box.

4. Shake in two further brood frames of bees.

5. Add at least two good frames of food and adhering bees, taken from the same colony.

6. Add frames of foundation or drawn comb to fill any gaps.

7. Close the nuc and place it next to, or on top of, the hive from which the frames have been removed. When the grass wilts, most of the older bees will return to the original hive. This is fine because the remaining, younger bees are more likely to accept a new queen.

8. Return after seven days and inspect the brood frames. There should be emergency queen cells where the bees make new queens from larvae of the right age. Remove anything that looks like a queen cell. The colony now has no way to make a new queen.

9. Study the brood frames and find a patch of capped brood that you think will emerge soon – ideally, somewhere where emerging bees are already nibbling through their wax cell cappings.

10. Shake or brush all the bees off this brood frame and take it indoors (although the following stages can also be done inside a shed, a car or in a large, transparent plastic bag).

11. Lie the brood frame on a flat surface near a window. Prepare the queen introduction cage by blocking the entrance chamber with fondant paste, or some marshmallow. Remove the plastic tab that allows access to the cage – this will enable bees in the nuc to slowly eat their way through to the queen. It's not uncommon for beekeepers to forget to remove the plastic tab, so double check that you have done so.

12. Open the travelling cage containing your new queen and let her crawl onto the brood frame. The attendant workers will come out as well. Remove them from the frame or let them fly to the window. The queen should stay on the brood, but because she is likely to be slimmer than usual after her journey she too could fly – which is why this procedure is best done indoors. Should the queen fly from the frame, let her land and then pick her up gently holding her by the thorax only, or gently scoop her back into the travelling cage and return her to the brood frame.

13. With the queen on some emerging brood on the brood frame, carefully lower the introduction cage over her, trapping her in place. Fix the cage to the brood frame using the supplied pins.

14. With the queen safely inside the introduction cage, carefully return the brood frame to the nuc and push the frames together. Close the nuc.

15. Leave the nuc for at least a week before checking that the queen has emerged from the cage, has been accepted by the colony and is laying eggs. If you cannot see queen or eggs, wait for another week. If you still cannot see evidence of the queen, give the colony a frame of eggs and young larvae from the main colony. If they make emergency queen cells you will know that the introduction has failed. If no queen cells are made, wait longer for the queen to start laying.

16. When the queen is laying well, unite the nucleus with the full-sized colony using the newspaper method.

1. Remove tab and fill the entrance with fondant.

2. Trap the queen on some emerging brood.

3. Return the brood frame to the hive.

4. Carefully push the frames together, sandwiching the cage between them.

Re-queening can be time-consuming and sometimes a little nerve-racking, but is a vital beekeeping skill. Done well, it ensures that your colonies are always in optimum condition and that your hobby is at its most rewarding.

Moving and uniting colonies

Moving bees

Among the many extraordinary talents of the honey bee is its ability to find its way back to the exact position of its colony after flights that might have taken it far from home. After spending the first three weeks of life as house bees, workers graduate to being foragers, flying from the nest to gather the resources that the colony needs to survive. Early flights are timid, hovering in front of the hive and facing the entrance. Gradually they become more adventurous, radiating outwards to observe and plot the position of local landmarks. With this knowledge, and using the sun as a navigational aid, foragers can fly far and then find their way home. In the final approach they will recognise the shape, size and colour of their hive before being guided to the entrance by the scent of the colony. Foragers can travel impressive distances, collecting nectar and pollen from perhaps 3 miles (5km) away. The distance will vary depending on where resources can be found. Urban bees typically forage within a few hundred metres of home, but their country cousins may travel much further.

This precision homing instinct can be useful to the beekeeper. During swarm management, for example, colonies can be divided into two hives and one of them

TIPS FOR MOVING A HIVE

1. Nucs are easily moved by one person. A full-sized hive will need two people to lift it.

2. Brood boxes should be full of frames when moved. Loose frames can slap together and crush bees.

3. Move a hive so that the frames inside are pointing in the direction of travel – this prevents frames from swinging too much.

4. Remove supers, queen excluders and any other equipment that will make the hive unnecessarily heavy.

5. Move hives in the early morning or evening when bees aren't flying and it is cooler (closed hives can overheat).

6. Close the door or block the entrance with sponge and block the feed hole in the crownboard so that bees cannot escape.

7. Secure the floor, brood box and crownboard together using ratchet straps. Use two straps in parallel to stop the hive parts scissoring during the move and creating gaps for the bees to escape.

8. A barrow or trolley of some sort can be used for short moves.

9. Make sure your route is clear and be sure to prepare the destination: can you park nearby, are the gates locked and are empty hive stands ready and waiting?

10. Wearing a bee suit is essential. Bees will find any small holes and escape during a move, so be ready to put your veil on, especially if driving.

11. Open the hive entrance as soon as it is in its new position – and be ready to retreat if the bees are agitated.

ABOVE Block the feed holes in the crownboard and strap the hive tightly together.

ABOVE Entrance sealed with foam.

moved to a different position. The older bees will leave the new site and fly back to the place they think of as home, achieving a division of old and young bees that helps to prevent swarming (see page 87). Another useful trick is to move the hive of a defensive, difficult-to-handle colony several metres to one side, placing a temporary hive or nuc box on the old site. The older, more troublesome, bees will fly back to the original position, leaving behind a reduced population that is easier to inspect and manage.

However, there are disadvantages too. Because bees know precisely where home is, it can be difficult to move colonies without planning and effort. The maxim to remember here is what beekeepers call the 'less than 3 feet or more than 3 miles' rule (the metric version of which would be 'less than 1m or more than 5km').

ABOVE Every wheelbarrow in the neighbourhood was borrowed to help move this apiary by one meter (3ft) a day.

The theory is simple. Because honey bees normally forage within a radius of approximately 1.5 miles (about 2.5 km) from their hive, moving a colony double this distance will mean foragers on long-distance flights are less likely to stray into known, internally mapped territory and return to the old hive position. On the other hand, moving a colony no more than 3 feet (1m) in any direction means that returning bees will still be able to find their home without too much difficulty. Much further and they are stumped, so moving colonies less than 3 feet or more than 3 miles is the rule of thumb.

Near and far

As we have seen, the simplest way to move a colony of bees without the possibility of foragers returning to their old home is to move them more than 3 miles. This means that if you have more than one apiary, it is useful if they are at least 3 miles apart so that colonies can be moved between them freely.

It gets more complicated when you want to move colonies short distances within an apiary. One way is to move a hive 3 miles away, leave it there for about two weeks for the bees to learn their new location, and then move it back to its new position just a stone's throw from where it began. Alternatively, the 'less than 3 feet' rule can be implemented to reposition colonies incrementally, moving them towards their new location in a series of short journeys, allowing the bees to reorient themselves to the new spot after each move.

If you move a hive backwards or forwards with the entrance orientation remaining the same, the bees will return on their normal flight path and find home slightly before or after they are expecting it. Sideways moves are more confusing for the bees, so make these shorter than forward or backward ones. Turning the orientation of a hive entrance makes it harder for bees to find home. So, if you are changing the entrance direction, turn the hive by no more than about 30° each time it is repositioned.

During good flying weather, a colony can be moved up to 3 feet (1m) every 24–48 hours. Each day the bees will reorient to the new position. But if the weather is cold and wet, wait until flying conditions improve so that the bees have a chance to fly and to learn the new position before the next move.

It's helpful to provide visual clues so that returning foragers can identify a newly positioned home. One method is to paint hives or landing boards different colours. Another is to pin different shapes to the front of hives. These can be coloured or have a distinct pattern. You could also position a large pot plant or other object next to the entrance – this signpost will also have to be repositioned with each move.

Tricking bees

Sometimes it's not possible to move a hive 3 miles – or convenient to move it in increments. For example, if moving a hive from a front garden to a back garden, it would be somewhat impractical and probably quite controversial for it to progress in stages through the house via the kitchen and living room. In such cases, several tricks can be used to persuade the bees to reorientate to their new position. Try facing a moved hive in a new direction, stuffing the entrance with fresh grass and covering the entrance with a dense pile

BELOW Covering the entrance of a moved hive can trick bees into accepting the new location as home.

of leafy branches (ivy is ideal). As the grass gradually wilts over a day or so, the bees will emerge from their hive to find an unexpected thicket of vegetation. After crawling through it, they will find the sun in the 'wrong' position and reorient themselves to the new location. A word of warning, though: this technique is not always 100% successful and you will sometimes find a handful of bees flying around the old location looking for home. Drones in particular often return to the old position.

Winter opportunities

In mid-winter, bees may not be able to leave their hive for weeks at a time, and then only for brief cleansing flights – they can only keep their legs crossed for so long. This gives an opportunity to move colonies any distance you like without worrying about the usual precautions. After confinement to the hive for so long, bees tend to reorient themselves on re-emergence, and their short cleansing flights help them to get used to the new location. Wait until there has been about a week of cold weather and at least another week of cold is forecast, and then move the colonies any distance you like. Move the hive gently in order not to disturb the bees' winter cluster.

Managing colony numbers

You might start beekeeping with a single colony, but by your second year you should aim to run a minimum of two colonies. One can then support the other, should you encounter problems. Colony numbers tend to increase naturally through the acquisition of swarms or by splitting colonies as part of swarm management, but you can also choose to make increase (creating more colonies) using various management techniques such as those mentioned earlier.

In fact, it is easy to suddenly find that you have too many colonies and that some downsizing is required. Your beekeeping might be taking up too much time, money or domestic space, or you might want to weed out colonies that are weak, unproductive or overly defensive. In such cases, there are ways to rationalise your stocks so that you have fewer, better colonies.

Most beekeepers soon work out the best number of colonies for their situation and beekeeping ambitions. The number of colonies kept then tends to fluctuate around the desired figure, usually increasing in spring and summer before being reduced again before winter.

Colony increase

When using the nucleus method of swarm management (see page 87), you divide a colony of bees wanting to swarm into two units, both having a good chance of survival and therefore doubling your stocks. However, the parent colony has the ability to produce multiple new queens, so there is opportunity to create more than one new colony.

To produce a three-way split when using the nucleus method of swarm management, proceed as before, removing the old queen to a nucleus box. The parent colony is then left with multiple open queen cells, ideally on several frames. When you return to the parent colony seven days later it can be split in half, half of the brood and stores staying in the original hive and half being transferred into another nucleus box. Each division should be left with a single, large, sealed queen cell. Because the flying bees – important for bringing home new stores of pollen and nectar – are oriented to the position of the parent hive, you must shake the bees from at least two frames of stores into the nuc. These frames will contain mostly older, flying bees. So that the flying bees do not return home, the newly created nuc is best taken at least 3 miles (5km) away. It can be brought home again in about a month, once the new queen is laying. Splits like this are best done early in the season so that colonies have time to build up before winter. Consider feeding split colonies with sugar syrup (see page 129) if there is a chance that they might run low on stores because of their smaller foraging force, or if there is bad weather.

A split of this type can also be made when a colony is not preparing to swarm, dividing it into two or more parts, one of which contains the old queen. As long as each of the queenless units has eggs or very young larvae on the day of the split, the workers will be able to make new queens under the emergency impulse (see page 94). If such a split is done early in the season, the colony with the old queen might still build sufficiently to bring in a crop of summer honey. By the time the new queens

have mated and started to lay, their colonies should be able to build up enough strength and stores to get through winter.

Uniting

If you have too many colonies, or some are weak, overly defensive or have an old queen, they can be combined, or united, joining a poor-quality colony with a good colony to produce a single, stronger colony that retains the characteristics that you want. If you have split colonies for swarm management purposes, you might find that either the colony with the old queen, or that with the new one, is unsatisfactory or surplus to requirements, in which case they can be united under just one of the queens. In autumn you might unite a weak colony with a strong one, to produce a single colony with a better chance of surviving winter.

You cannot mix and match bees from different colonies and expect them to integrate harmoniously. Every colony has distinct characteristics, including a unique colony odour resulting from the various pheromones produced by both worker bees and the queen. Mixing bees from two colonies without taking sufficient precautions will result in fighting and deaths.

The most reliable way to unite colonies harmoniously is called the newspaper method. This is best done in late summer or early autumn, when there are no heavy supers on the hives.

BELOW Newspaper and queen excluder over broodbox, ready to accept second colony.

ABOVE Two colonies in the process of uniting.

ABOVE After a few days, the newspaper is chewed through and the colonies mingle.

If two colonies to be combined are some distance apart, gradually move the weaker colony towards the stronger one. Every few days, lift and carry the hive no more than one metre at a time until it is next to the strong colony. This way the flying bees will continually re-orientate to the new position of their hive, rather than flying back to the old site. If it is a full-sized hive with lots of frames, you might need some help with the lifting – or it can sit in a wheelbarrow and be incrementally pushed to its new position.

When the two colonies are sitting next to one another, prepare them for uniting, as follows.

1. Both colonies need to be in the same kind of hive. If one is in a nucleus box, move it to one side and place a floor and brood box in the same position. Then transfer the frames from the nucleus colony into the brood box before adding crownboard and roof.

2. Next, find the queen in the moved colony and kill her. This is never a pleasant job, but there is nothing else to be done with an unwanted queen. The quickest and most humane method is to pinch the queen's thorax between your thumb and forefinger. It takes only a second. Some people prefer to put queens into the freezer where they fall asleep and die slowly. This might be easier for the beekeeper, but I am not sure it is any better for the queen.

3. Prepare the strong colony to receive the weak one. Remove the roof and crownboard to reveal the top bars in the brood box. Scrape off any lumps of comb. Lay one or two sheets of newspaper across the top of the frames so that there are no gaps. Place a queen excluder over the newspaper to hold it in place.

4. Now place the brood box of the weaker colony on top of the queen excluder. Place the crownboard and roof on top of your double-decker hive to complete the operation.

Over the next few days, bees from both colonies will nibble through the newspaper barrier, gradually acclimatising to one another as they do so. You will find finely shredded newspaper under the floor or outside the entrance. About one week later, move the top brood box to one side and then go through both brood boxes, selecting the best frames of brood and food to be combined in the bottom brood box. Remember that the queen you have chosen to keep is in the bottom brood box, so it is worth finding her and putting her and the frame she is on in a spare box for safe keeping during the operation – having deliberately killed one queen, you don't want to accidentally lose the other. When the frames in the bottom brood box are arranged, reintroduce the frame with the queen on it. You have now united two colonies to produce a single new one.

SOME GENERAL RULES ABOUT UNITING COLONIES

- Both colonies should be healthy and free of disease.
- At least one of the colonies to be united should be strong – uniting two weak colonies rarely produces a strong one.
- Move the weak colony to the strong colony and place the weaker one on top.
- The queenless colony generally goes on top.

ABOVE Two colonies have been united, and several brood boxes are being used to sort out the best frames for use in the consolidated hive.

BELOW A good-sized colony of honey bees actively flying on a warm day.

The rewards of beekeeping

Forage

Bees over the age of about three weeks regularly leave the nest to collect the resources that the colony needs to survive and grow. Beekeepers call these bees foragers, and refer to what they collect as forage. One of the great delights of honey bees, and what makes them different to any other form of livestock, is that – for the most part – they feed themselves; they are free range in the truest sense of the word. Honey bees will fly perhaps 3 miles (5km) from their hive to find and collect what they need, meaning they have the potential to exploit thousands of acres of other people's gardens, farmland and woods. No wonder a beekeeper wrote the following (unattributed) rhyme:

> *God bless the bee.*
> *Long may she yield*
> *A harvest from my neighbour's field!*

LEFT Nectar-filled cells at the height of a flow.

RIGHT A selection of pollen grains under the microscope.

BELOW With experience it is possible to identify pollens carried on the legs of bees. Some are quite distinct although many are unhelpfully similar shades of yellow.

Pollen

Pollen provides honey bee colonies with proteins, fats and minerals. It is used in the production of brood food, which worker bees feed to larvae. It takes about 125mg of pollen to produce each new bee. This might not sound like much, but it means that the average colony must collect around 45kg (100lb) of pollen a year – impressive when you consider that pollen is little more than dust.

Using the techniques described earlier (see page 28) a honey bee forager collects pollen from flowers and brings it back to the hive. There, she scrapes the pollen off her back legs into a cell, usually near the brood nest.

ABOVE A worker carrying the distinctive red pollen of horse chestnut.

BELOW A rainbow of pollens stored for later use.

Using her mandibles, she packs the pollen neatly into the cell, compressing it to make room for more. When a cell is filled, a little honey is added to the pollen and bacteria begin to grow on it, producing lactic acid and reducing the pH of the pollen to increase its storage life. This pickled pollen is known as bee bread. Nurse bees consume the stored pollen, which allows the glands in their bodies to produce brood food.

The nutritional content of pollens from different plants varies, so bees will visit a range of flowers to ensure a balanced diet. Bees with access to a limited variety of pollen, such as those on agricultural land where one flowering crop is grown on a large scale, can suffer nutritional deficiencies.

One of the great pleasures of beekeeping is watching your bees arriving at the hive entrance with variously coloured pollen loads. It is fascinating to decipher what flowers have been visited according to the colour of the pollen collected – although the colour of a flower is not necessarily the same as the colour of its pollen. The majority of pollens are a yellowish shade, but they can be orange, white, brown, green, blue or even black. You will learn to recognise some colours as an indicator that certain plants are flowering. The red pollen of horse chestnut, for example, is highly distinctive.

Propolis

Propolis is a gum-like substance collected by honey bees from plants and trees that exude sticky resins from their leaves, buds and branches. Foragers gather it with their mandibles and bring it back to the hive in their pollen baskets. Other bees then need to tease it out, before it can be used. Only a small percentage of foragers collect propolis, and it is quite rare to see bees in the hive with propolis on their legs. Nevertheless, some colonies collect it in substantial quantities.

One use of propolis is as a general-purpose building material, for filling up cracks, crevices and corners to make a hive weatherproof and keep out pests. It is also used to strengthen the edges of wax comb. Some colonies use it to build walls across the hive entrance, leaving only a small gap through which to enter and exit. The smaller gap keeps out the worst of the elements and is easier to defend against predators – it can be likened to a city wall. This explains the Greek origin of the word propolis – *pro* meaning before, and *polis* meaning city.

Propolis also has antimicrobial and antibacterial properties, and is used around the hive as a cleaning agent. Dead mice, snails and other potential pollutants that bees are unable to remove from the nest will be sealed with propolis to limit contamination. Colonies living in tree cavities line the walls almost entirely with propolis, producing a kind of protective envelope to contain the colony. Workers also apply a thin layer of propolis to the inside of cells where they want a queen to lay eggs.

Research has shown that colonies with more propolis in their nest are more resistant to diseases such as AFB (see page 140). Nevertheless, beekeepers often remove it because it makes inspecting hives hard work – the

ABOVE A sticky fate awaits those who unwisely enter the hive; a snail entombed in propolis.

RIGHT TOP Sticky propolis on the hind legs of a worker.

RIGHT A worker collecting water from a drop of dew.

caramel-like substance sticks hive parts, fingers and tools together and stains clothes. Propolis produces a rich, warm aroma in the hive and anywhere that beekeeping equipment is stored. Despite being something of a sticky annoyance, it is one of the sensual delights of beekeeping. Propolis harvested on a commercial scale is used to produce a range of effective medical products for humans, such as throat lozenges, tinctures and even a treatment for cancer.

Water

Honey bees use a surprising amount of water – up to a litre a day per colony in summer, according to some estimates. Water is used to cool hives in hot weather. Droplets are hung around the nest to absorb heat, and the bees then fan their wings *en masse* to create a draught, which evaporates the warm water and expels it from the hive.

Water is also used to liquify or dilute honey so that it can be consumed by the bees – stored honey is too concentrated for them to use as it is. In winter and spring it is especially important that honey bees can find a source of water near to their colonies for this purpose. Colonies can starve if cold weather prevents them from leaving the hive to collect water to dilute honey that has crystallised in the comb. This is especially true for

the particularly hard honey produced from ivy nectar, which is often gathered in the autumn and can form a substantial proportion of winter stores.

Bees will collect water from damp grass, puddles, garden ponds or even a dripping tap. When lots of water is required, many bees will visit a source and can be a nuisance to neighbours, with dozens of bees gathering at their bird bath or children's paddling pool. Try to provide a water source within your own garden – though not too near the hives as bees don't always forage for water close to home. Create a supply of shallow water with an area of mud, moss or leaves where bees can perch to suck up water without drowning. This could be at the edge of a garden pond, or even just a bucket of water with some sticks or stones in it. Put it somewhere that catches the sun so that the water can warm a little on cold days. Bees drinking very cold water can sometimes chill and die. Bees can be encouraged to visit a site in your garden by initially adding a little sugar to the water. Once they start using a water source, they will remain faithful to it unless it dries up, so keep it topped up in hot weather.

Nectar

Nectar is a sugar-rich liquid that is secreted by flowers from glandular structures called nectaries, which are usually found at the bottoms of the flowers. Honey bees collect nectar and convert it into honey, which they store and consume to give them energy. Although its primary importance is as a source of carbohydrates, nectar also contains small quantities of amino acids, proteins, ions and various microbial compounds that recent research suggests bees might use for self-medication.

The quantity and quality of nectar produced by different types of flowers vary. Some produce nectar with higher concentrations of sugar, or with a differing balance of sugar types. Honey produced from the nectar of different flowers can vary greatly in colour and flavour, and it is these differing qualities that make honey so appealing as a food to humans.

With a few exceptions, nectar is produced by plants only when they are in flower. This encourages visits from bees and other pollinators who will transfer pollen from one flower to the next as they collect nectar, thereby causing fertilisation and allowing the formation of fruits, nuts and seeds.

Although different flowers produce varying amounts of nectar, when climatic and environmental conditions are just right flowers can produce much more than at other times. Beekeepers call this a flow. During a flow, a constant stream of foragers flies to and from the hive and there is a tremendous sense of industry. If you inspect a hive during a flow the bees will pay you barely any attention. At night, colonies hum to the sound of nectar being processed into honey. Hives can gain as much as 5kg in weight in a single day. Flows can last for a day or two or for several weeks, and can come to a very sudden end, indicated by a noticeable drop in bee activity.

There can be several flows throughout the beekeeping season, with hives getting heavier during a flow and lighter again between flows as some of the honey is consumed. Broadly, there are two main flows in a year, each encompassing smaller flows from various seasonally flowering plants and all dependent on the weather. In the southern UK, the spring flow typically lasts from April to early June and includes nectar from trees such as apple, cherry, maple and hawthorn, as well as dandelions and fields of bright yellow oilseed rape. After this comes the summer flow, typically lasting from mid-June to late July. This includes nectar from summer

LEFT AND BELOW White clover is a significant source of UK honey, although conditions need to be just right to provide a flow.

flowers like white clover and bramble, trees such as lime, and agricultural crops, including field beans and borage, as well as lesser contributions from hundreds of wild and cultivated flowering plants. In some years in the UK there is a very significant drop in available nectar between the spring and summer flows. This is called the June gap, and honey bee colonies can starve at this time (see page 131). In some years, and depending on location, there can be substantial late-summer flows from heather, ivy and Himalayan balsam.

In the early part of the year, incoming nectar is used to produce young bees and build the colony. Later, the bees concentrate on storing honey for winter use. Beekeepers who want to harvest honey hope that by the end of the summer flow, their colonies will have stored enough for the bees to survive winter, allowing any surplus to be harvested.

Pollen analysis of a typical UK honey might show that bees have visited the flowers of 30–50 species of plant. Nectar from all of these may contribute to the honey, but in most cases the majority of honey probably results from a few abundant, nectar-rich species and can be highly dependent upon location. In rural areas, oilseed rape and field beans can be important sources. In towns, trees are commonly the most significant source. Some crops will be very localised, depending on the presence in a single area of a particularly productive crop or wildflower. Heather honey only occurs where there are large areas of the plant on moors or heaths. Overall, the single most important UK honey flower in terms of volume is probably bramble, followed by ivy, which is rarely harvested by beekeepers.

Other sources of nectar

Not all nectar comes from flowers. Some plants have extrafloral nectaries – small areas, usually on stalks or the underside of leaves, which secrete nectar that can be gathered by bees. Broad beans, or field beans, have extrafloral nectaries on small leaf-like outgrowths called stipules at stem junctions. These secrete nectar before the plant begins to flower. When grown on an agricultural scale, these can produce a large crop of very tasty honey. Other sources of extrafloral nectar include cherry laurel, elderberry, passionflower and even bracken. In Asia, large quantities of honey come from the extrafloral nectaries of the rubber tree.

ABOVE Honey bee visiting the extrafloral nectaries on the underside of cherry laurel leaves.

ABOVE Honeydew on a beech leaf.

Bees also gather a sugary substance that is produced by the metabolism of other creatures. Aphids and scale insects tap into the sap of the plants and trees that they live on, sucking up large quantities of liquid which they process to extract the small amount of protein it contains. They are left with an excess of sugar which they secrete from their rear. This sticky substance can cover leaves and even drip to pavements and cars. In the morning the sugar is diluted by dew, making it liquid enough for bees to suck up. The result is honeydew honey, a dark, strongly flavoured kind of honey. Most honeydew honey comes from trees such as oak, plum, fir and beech, and only in years when there are heavy aphid infestations or a lack of other nectars for bees to collect. In the UK it can add to a summer crop of honey, but in highly forested areas, such as in Germany, Greece and Turkey, honeydew honey can be a major crop in itself, commonly called forest honey.

The top UK plants for honey production

SPRING

Hawthorn

Oilseed rape

Sycamore

Dandelion

Tree fruit (apple, plum, pear, cherry)

SUMMER

Field beans

White clover

Bramble

Lime trees

Sweet chestnut

AUTUMN

Heather

Ivy

Himalayan balsam

Finding forage and making honey

The way that bees harvest and process what they forage is fascinating, but the way that they communicate with one another about their finds and organise their foraging trips is truly staggering. A small proportion of older flying bees operate as scout bees. After searching a wide area they return with information about the quantity, quality and location of what they have found, and recruit other foragers to exploit their discoveries on a large scale. This information is encoded in dances – a sophisticated bee language that we can 'read', thanks initially to the Nobel Prize-winning work of Professor Karl von Frisch in the 1920s.

The waggle dance

A scout bee returning with information about a source of forage dances in a figure-of-eight movement on the surface of the comb. The central axis of her dance is at an angle from vertical (0°). Recruits reading the dance know that vertical represents the position of the sun and that the angle of the scout bee's dance represents the number of degrees from the sun that they should fly. If the scout has found forage at 20° to the right of the sun, she will run on a line 20° from vertical inside the hive. When the forager exits the hive, she will identify the position of the sun and fly 20° to the right of it. Extraordinarily, the scout bee will slightly alter the directional information that she gives, to account for the fact that the sun's position in the sky will have changed since her initial observations. Furthermore, the dance is done in the darkness of the hive; recruits cannot see it but must sense it through touch and vibrations.

The waggle dance communicates more than simply direction of travel. As the scout runs along the central axis of her dance route, she waggles her abdomen. The vigour of the waggle indicates how rich the food source is. The frequency with which the scout runs along the central axis indicates the distance that foragers need to fly to find the forage being advertised. By licking

ABOVE A scout bee (with pollen on her head) dancing for her recruits.

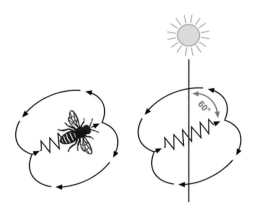

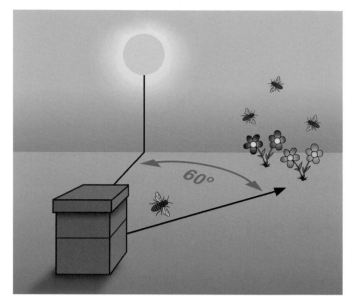

ABOVE AND RIGHT By dancing at an angle from vertical, scout bees give recruits directional and distance information to guide them towards sources of forage.

(palpating) the scout's abdomen, recruited foragers derive information about the type of flowers they are looking for, the waxy coating having absorbed their scent. To reinforce this, the scout feeds them samples of the nectar so that they know exactly what they are seeking.

Until recently, it was thought that there were two different kinds of dance; a 'round' dance indicating that there is forage to be sought in the general nearby area and a 'waggle' dance giving more detailed information about more distant finds. In fact, modern high-speed photography reveals that there is only one dance form using the same techniques – it just appears somewhat different to the human viewer.

Continuing research suggests that a variety of other methods are used to indicate how and where to locate forage, demonstrating that honey bees, like so many other creatures, operate in ways more subtle and refined than we have historically given them credit for.

How bees make honey

At a basic level, nectar is water with a small amount of sugar dissolved in it. The water to sugar ratio differs between plant species, with most nectars containing between 10–40 per cent sugar. Because they try to work efficiently, honey bees will favour visiting flowers that offer richer nectar, even flying much further to collect it as long as the end result is more profitable in terms of sugar gained relative to energy expended.

Foragers visiting flowers use their proboscis, an extendable straw-like appendage to their mouth, to transport nectar into their body. They may suck up the nectar directly or lap it up with their spongy tongue. Turning this nectar into honey is a two-stage process that begins inside the body of the bee. In the first stage, enzymes produced in the hypopharyngeal glands, located in the head of the bee, are added to the nectar as it travels to the honey stomach, or crop, where it is stored during transportation. The enzymes make chemical changes to the sugars in the nectar, converting disaccharides into simpler monosaccharides. More technically, this means splitting each molecule of sucrose into two smaller molecules – one of glucose and one of fructose. Some of the glucose is further broken down to create gluconic acid and hydrogen peroxide, which are important for destroying bacteria and preserving the honey.

Arriving back at the hive, foragers regurgitate the nectar, passing it to house bees (called receiver bees) which swallow it, add more enzymes, and take it into the hive for storage before further processing. Meanwhile, the forager that collected the nectar immediately returns to the field to find more.

Nectar arriving at the nest has already started its chemical conversion into honey by having its sugars

ABOVE Forager using her proboscis to reach the nectaries in apple blossom.

altered with enzymes. However, it is still too liquid to be safely and efficiently stored for the long term. The second stage of the process is to remove much of the water, turning it into the thick, sticky substance we know as honey. This is done through evaporation. Nectar is sucked up from wherever it is stored by workers who hang droplets of it on their probosces. There it is rolled up and down to continually expose the surface of the nectar to the high temperatures in the hive. At night, when all the workers are in the hive, many thousands of them will do this work. Others will fan their wings, drawing air into the hive and passing it over the nectar, evaporating off excess moisture. The moist air is then directed out of the hive entrance by more fanning bees. In this way, the moisture in the nectar is removed, until it reaches a water content of around 18 per cent. This finished honey is redistributed within the hive to wherever the bees want to store it, either around the brood nest or in supers if there are any on the hive. If they can, bees prefer to store honey above brood. Once a cell is filled to the brim with honey, a wax cap is built over the top, keeping the honey in perfect condition until it is needed.

The statistics involved in honey production can boggle the mind, and possibly prick the conscience as you spread it on your toast. While foraging, a worker may fill her crop with approximately 40 mg of nectar – a mere droplet. The nectar, mostly water, will be reduced by about one-third by the time it has been condensed into honey. A one-pound (454g) jar of honey may require around two million flower visits and 50,000 miles (80,000km) of flying to produce – so it can hardly be said to be a food with low air miles!

ABOVE Forager passing nectar to a receiver bee. A process called trophallaxis.

The honey crop

A healthy honey bee colony will survive the winter by feeding on the honey that it made and stored during the previous summer and autumn. By late winter, much of the comb in the brood nest will have been emptied of stores. When the first significant trees and flowers bloom in early to mid-spring, foragers will begin bringing home fresh supplies of nectar and pollen. At first, this will be sufficient only to raise the young from the small number of eggs laid by the queen at this time of year. As the weather improves and more flowers emerge, more nectar will be collected and stored, being used to raise increasing numbers of young as well as replacing stores consumed over winter. At some stage, and if the weather remains fair, much of the comb in the brood box will again be filled with honey. Freshly collected nectar must then be stored wherever space can be found for it. Often this will be in empty cells in the brood nest, which means there will be less space for the queen

to lay eggs. This can impede colony growth and may trigger swarming preparations. To avoid this, beekeepers give colonies more space, expanding hives using supers – additional, shallow boxes where the bees can store nectar and process it into honey. Honey stored in supers is generally considered surplus to the requirements of the colony and can therefore be harvested by beekeepers. A queen excluder is added over the brood box, preventing the queen from moving into the super above and laying eggs where only honey is wanted.

Adding supers

When in spring to add the first super will depend on the weather, your location, and how the colony is developing. If you have already kept bees for more than one year, you can use a super containing frames of comb that the bees made in previous years. These will have had the honey extracted from them before being stored over winter. The bees will quickly repair this pre-made comb and begin making use of it. New beekeepers without frames of

ABOVE Once nectar has been converted into honey the cells are sealed with wax.

ABOVE Adding a super of new frames and foundation.

comb will have to add a super containing frames of wax foundation. This still has to be drawn into comb by the bees, which means it should be added earlier.

Add the first super as soon as the brood box appears to be full of bees – in other words when, in about mid-spring, bees are working on every frame. In the UK, supers are usually added when wild cherry, blackthorn and, in agricultural areas, oilseed rape come into flower. Keep an eye on the frames in a super; when bees are covering and working on about three-quarters of them, with honey stored on the central half dozen or so, consider adding a second super to give even more space. Remember that when nectar is first brought into the hive it occupies about three times the space that it will once it has been condensed into honey – it all has to go somewhere. Supers are added incrementally because adding too many at once creates a large space above the brood nest, which in spring is still at risk of getting chilled, particularly at night. It can also result in chimneying, where the bees build comb and store honey only in frames above one another, rather than first working on the frames to either side. You will need two or three supers for each productive colony.

Spring honey

In the UK, most areas have two main honey flows a year. Each can result in a harvest of honey for the beekeeper. The spring flow typically begins in mid-spring and lasts until early summer. In many areas the nectar gathered by bees comes almost entirely from trees, including maple, sycamore, cherry, apple and hawthorn. Dandelions can also give much nectar at this time. In rural areas the biggest source of nectar can be fields of oilseed rape. In the right conditions, rape can provide a huge crop of light, mild-tasting honey. Importantly, the nectar and pollen supplied by this crop also provides a massive boost to brood-rearing, which means there will be many more bees available to exploit the summer honey flow when it comes in about another two months. If you have rape in your area, it will dictate when you harvest your spring honey. It needs to be removed from the hive before it crystallises and cannot easily be extracted from the comb. As a rule of thumb, spring honey that includes rape should be harvested as soon as the fields of rape start to look more green than yellow.

Summer honey

In some years the spring and summer flows run seamlessly from one to another. In other years there is a distinct and sometimes prolonged gap between flows – often called the June gap because of its typical timing in southern Britain. This can be a precarious time, when large colonies of bees sometimes starve, particularly if the beekeeper has removed all of the spring honey and there is no fresh nectar to be gathered.

117

The summer flow is the main event when it comes to honey production. Lime trees, white clover and bramble are usually the most significant nectar providers, but hundreds of wild and cultivated flowers will also contribute. In agricultural areas, field beans flowering in the early part of the season can provide much nectar. The flow can be continuous or proceed in fits and starts according to the weather. It is important to continue checking colonies on a weekly basis, both to keep an eye on possible swarming and to monitor whether additional supers need to be added. In the UK, one or two supers full of honey per hive is about average, but a strong colony in a good year can produce two or three times as much. The summer flow is usually over and the honey ready for harvesting by about mid-summer, which would typically be towards the end of July in southern Britain, although this varies from year to year.

In some areas it is possible to get a late summer crop of heather honey. Two types of heather can

LEFT Hawthorn can be a significant source of pollen and nectar in spring.

ABOVE A well-planted summer border can be a long-lasting buffet for bees.

produce a crop in the UK. First to flower are bell heathers (*Erica cinerea*, and also *E. vagans* in small areas of Cornwall – strictly speaking these are heaths rather than heathers). These produce a very distinctive-tasting honey the colour of port wine. Later comes ling heather (*Calluna vulgaris*), which produces a jelly-like honey that is thixotrophic, meaning it will not flow unless agitated. Producing heather honey is a specialist pursuit that usually involves transporting colonies to remote moor locations for the flowering period, as well as the use of dedicated extraction equipment.

Harvesting honey

Honey should be harvested only when it is ripe – that is, when the bees have finished converting nectar into honey by removing most of the water. The UK legal maximum water content of blossom honey for sale is 20 per cent, but around 18 per cent is a better target for the beekeeper. This is because with a water content above 20 per cent it is possible for certain bacteria and yeasts to live in the honey and cause fermentation. Honey with a water content of 18 per cent or less, sealed in the comb or in a jar, will remain in good condition for many years.

Harvest spring and summer honey when you think each flow has stopped. Sometimes it is obvious from the lack of flying activity at the hive entrance that little nectar is being gathered, but keep an eye on frames of honey in the supers to see if more nectar is being brought in. Even after a flow has finished, bees continue to ripen and reorganise the storage of honey for some days, so wait until it looks as if no more honey is being capped before starting to harvest it.

Generally, only frames with at least 75 per cent of their honey cells capped should be harvested. When mixed together, the honey from such frames should have the correct water content. Sometimes bees take a long time to cap honey even when it is ripe. Rather than waiting for capping, you can apply a simple test; lift a honey frame from a super, hold it horizontally so that the uncapped cells face downwards, and give it a sharp shake. If nothing more than the odd spot of honey flies out, it should be safe to harvest. A better method is to test with a refractometer. These are reliable, easy to use and reasonably priced. A dab of honey placed between two glass surfaces can have its water content read through an eyepiece.

ABOVE A worker foraging from late-summer ling heather on Exmoor.

ABOVE A perfect frame of ripe, sealed honey, ready for harvesting.

Clearing supers

To harvest honey, the bees first need to be removed from the frames. There are two ways of doing this. Bee escapes are one-way doors that allow bees to exit supers and not return. The escapes can be temporarily added to the hive's usual crownboard, or permanently built into what is called a clearer board. Most hives come supplied with Porter bee escapes which can be fitted into the slots in the crownboard. Much more effective are the larger escapes such as those pictured below. The afternoon before harvesting, place a board fitted with escapes between the brood box and the supers. Make sure the escapes are the right way round so that bees can move down from supers to brood box, not the other way around. By the following morning, the majority of the bees will have moved into the brood box. The supers can then have any stragglers brushed away before being taken away for extraction.

Alternatively, bees can be removed manually, one frame at a time. Remove the supers from the hive then take one honey frame at a time and shake the adhering bees back into the hive. Use a bee brush to gently brush any remaining bees from the comb back into the hive. Place the cleared frame in an empty super and cover it with a cloth to prevent bees from returning. Better still, store it in a lidded plastic box which can be used to transport honey frames to and from your extracting area with minimal dribbling of honey.

Once frames are removed from the hive and cleared of bees, they should be processed as quickly as possible, while the honey remains warm and is easier to extract.

BELOW The underside of a clearer board – bees have been funnelled down, out of the super above.

Extracting honey

Honey extraction can be a time-consuming and messy process. Having the appropriate equipment and being organised about how you use it makes life much easier. Unless you are fortunate enough to have a dedicated honey room (and some beekeepers do), your extracting will probably be done in the household kitchen.

As you will be dealing with a food product, hygiene is an important consideration. Before extracting begins, wash all surfaces and the floor, and remove any pets and lingering hairs. You will need at least two sources of water – one for washing your hands, and another for washing tools and equipment. A kitchen sink is fine for hands, and a separate bowl of water with a sponge or cloth can be used for tools. It's a good idea to have a bucket of water and a cloth or mop ready for clearing up spillages. Wipe up spilled honey with a damp cloth rather than a soaking wet one which will only spread sticky patches further. Use cold water, not hot. A supply of clean tea towels is useful for drying hands and equipment. It's a good idea to cover the floor with newspaper or a couple of inexpensive shower curtains that can go into the washing machine afterwards; however careful you are, drips of honey will land on the floor and can easily be transferred around the kitchen and then the whole house. Keep external doors and windows shut during extraction; if one honey bee finds its way in, there can soon be many more. Containers and tools used for honey extraction and storage should be made of food-grade materials.

Design your workflow

It's easy to get in a messy muddle when extracting. Arranging your workspace and designing a workflow helps to keep things organised, efficient and clean. In order of use, you will need: somewhere to place supers or boxes of honey frames; an area for uncapping frames; nearby floor space for the extractor; somewhere to place extracted frames ready to return to the hive (which can be the box or super they started off in); and somewhere for buckets of extracted honey. This all requires quite a lot of space but can be managed in most domestic kitchens, perhaps with a little reorganisation, suspension of normal activities and an understanding family. Ideally, each stage should be arranged in a close circle around where you will stand, avoiding any great distances over which frames can drip sticky honey during transfer.

Uncapping

Before extraction, honeycomb needs to have its wax cappings removed. There are several ways to do this. A sharp knife (either a dedicated uncapping knife or your scrupulously cleaned kitchen bread knife) can be used to slice off the cappings. Alternatively, an uncapping fork can be used to pick away the cappings, or a roller, rather like a spiky paint roller, can be used to pierce holes in them. The cleverly designed uncapping slice is a tool that can be dragged across the surface of the comb to razor off and remove wax cappings, and is perhaps the most efficient method. Most people end up using a combination of the above as different frames can present different challenges.

Uncapping should be done over a container to catch removed wax and dripping honey. The best option is a commercially available uncapping tray. This is a large plastic box, with a perforated floor about halfway up. The lug at one end of a frame is fitted into a slot while the lug at the other end is held by the operator, leaving one hand free to slice, pierce or fork the cappings using the chosen tool. Removed wax falls into the box below, and honey drips through the perforated floor into a chamber beneath, from which it can be drained via a tap. Various designs of uncapping tray are available to buy or they can be homemade – but make sure any container you use is made of food-grade plastic. When both sides of a frame have been uncapped, it goes into the extractor.

Extractors

Extractors use centrifugal force to throw the honey out of the uncapped cells. The honey frames are arranged in a cage which is spun around a central axle, either by turning a handle or with the help of an electric motor. An extractor is probably the most expensive piece of beekeeping equipment you will buy. If you join your local beekeeping association it is usually possible to hire one for a very reasonable price, the only problem usually being that everybody wants to use it at the same time. Broadly, there are two types of extractor: tangential and radial.

ABOVE Uncapping with a knife.

ABOVE An uncapping slice.

ABOVE An uncapping fork.

ABOVE An uncapping tray. It can get rather messy.

Tangential extractors

These are cheaper and more often used by beginners. They have a square or triangular mesh cage which spins around a central axle and are generally hand-powered. Most models take between two and six frames of honey at a time. The uncapped frames are placed into the cage

ABOVE Uncapped frame in a tangential extractor.

ABOVE Loading a radial extractor.

ABOVE Spinning frames in a radial extractor.

with one side facing the outer barrel of the extractor and one facing inwards. When the cage is spun, the honey in the outer face of the comb is flung onto the side wall of the extractor, and runs down to a sump at the bottom from where it can be emptied via a tap. Because the honey is removed only from one side of a frame at a time, the frames have to be turned over several times to extract all of the honey and ensure that the heavy honey in the cells of the inward-facing comb doesn't push through and destroy the weaker, emptied cells of the outward-facing comb. Repeatedly turning the frames is a messy and time-consuming business, but this method is the most effective for removing the majority of honey from a frame.

Radial extractors

These have a circular cage that allows frames of honey to be arranged around the axle like the spokes of a wheel. This allows more frames to be extracted at once– anything from six to more than 100 in the case of commercial versions. Honey is removed from both sides of a frame at the same time, so they don't need to be turned over throughout the process, saving a lot of time and mess. Extractors can be hand- or motor-driven. The great advantage of motor-driven extractors is that they can be left spinning while you uncap the next batch of frames, making the process much more efficient. If you use a motor, make sure the spinning speed starts slowly and is gradually increased as the combs start to empty of honey.

Loading extractors

Because the cage in an extractor rotates around a central point, it is important to load frames in a balanced way. Frames of honey can be different weights, so try to place ones of approximately the same weight opposite each other. Extractors with an unbalanced load can wobble and rock alarmingly, making them hard to operate and placing great strain on the components. Some extractors come with a built-in stand, or they can be fixed to a homemade frame. Fitting castors to the feet allows unbalanced extractors to jiggle around during operation but without the dramatic and potentially harmful shaking. It's tempting and enjoyable to watch the honey flying out of the frames during extraction, but keeping the lid shut creates a flow of air that better draws the honey from the comb.

Straining honey

Extracted honey drains to the sump at the bottom of the extractor, from where it can be drawn off through a tap. A double honey strainer, which is a sieve with two grades of mesh, can be used to remove most unwanted particles from the honey. Place the strainer over a honey bucket and sit it under the tap of the extractor. Let enough honey in to fill the strainer, and allow it to drain through before adding more. After a while the strainer will become clogged and you will need to clean it to maintain an efficient flow. Never leave the extractor tap open and the honey draining while you do another task, even for a moment or two – you are bound to forget and return to a puddle of honey creeping across the floor. Every beekeeper has a honey horror story about this kind of situation.

Jarring honey

Strained honey can be jarred and is fine for home consumption, but needs further treatment if you wish to sell it. Honey for immediate sale can be poured into

ABOVE Draining honey into a strainer. Who could resist dabbing in a finger to sample the harvest?

a settling tank. These hold a large amount of honey, and some have a strainer at the top to additionally filter honey as it is poured in. As the honey settles, bubbles and wax to float to the top and particles of dirt to sink to the bottom. After at least 24 hours the tap can be used to fill one jar at a time with beautifully clear, clean honey.

Carefully stored, honey will remain in perfect condition for years. Honey that you wish to sell in the future is best stored in bulk using food-grade plastic buckets and then prepared and jarred shortly before sale. After extraction, honey that has passed through a double strainer goes into these buckets and is tested with a refractometer to check that excessive moisture won't cause fermentation during storage (see page 119). The buckets should be filled to the rim so that no air is trapped under the air-tight lid. Store them somewhere dark and cool, such as in a garage.

Harvesting without an extractor

If you think you might only harvest a small amount of honey, or don't want the bother and expense of buying and maintaining extracting equipment, there are alternatives. Rather than providing bees with super frames fitted with wired wax foundation, frames with very thin unwired foundation can be used instead. This foundation is so thin that it can be eaten, along with the resulting wax comb and the honey stored in it. Alternatively, unwired foundation can be cut into 'starter' strips around 3cm (1inch) wide and fitted into the top of frames. The bees will draw out these strips into comb that fills the whole frame. When filled with capped honey, the entire slab is cut from the frame and further divided

ABOVE Extracted honey comb.

into smaller portions, before being packed in containers for storage or sale. Honey prepared in this way is called cut comb honey and is considered a delicacy by some. Conveniently, it can be stored frozen until required.

Alternatively, comb cut out of frames can be mashed and then strained through a sieve over a bowl to separate liquid honey from wax.

Rich rewards?

How much honey can you expect to harvest from your hives? Every colony is different; every location is different; every season is different. If everything comes together perfectly for two glorious weeks in spring, and again in mid-summer, you could get a bumper crop. In the UK that might mean 30–40kg (66–88lb) per colony, but the average UK crop is about 11kg (24lb). Some years are much better than others, but in some there will be no honey at all. Don't give up the day job just yet.

Extracted frames

After extraction, return the supers and emptied frames to the hives that they have come from. Do this at dusk to prevent bees from different colonies trying to rob one another. Leave the frames in place for several days – the bees will lick them bone dry. Then remove all but one super of the cleaned frames, leaving it in place for the next honey flow. Add further supers as needed. Extracted frames can be stored over winter in plastic boxes or stacked supers, protecting them from wax moth if necessary (see page 150).

ABOVE Winter storage for extracted frames.

Storing and selling honey

Honey can remain in perfect condition for a long time when stored correctly. However, it can be subject to various changes that the beekeeper must be aware of in order to guarantee a high-quality jar of honey for home use or sale.

Crystallisation

Crystallisation, or granulation, is the process of liquid honey turning into a solid. Supermarket honey usually stays liquid for a long time because it has been processed in a way that stops or slows crystallisation – often at the expense of the finer qualities of the honey. Honey harvested from your own bees will usually crystallise in anything from a few days to a few months, depending mostly on the type of flowers that have contributed their nectar to make it.

Honey is a complex mixture of various liquid sugars, mainly fructose and glucose. Glucose quickly comes out of solution, forming small crystals around which other crystals soon form, turning the honey hard. Some nectars contain higher levels of glucose than others and will crystallise more rapidly. Oilseed rape nectar has such high levels of glucose that the honey it makes can set hard while it is still in the comb, or even in the extractor during processing. If allowed to set naturally, such honey can be very hard and almost impossible to remove from a jar with a spoon. However, by carefully controlling the way honey crystallises, it is possible to

ABOVE Honey beginning to crystallise in the jar.

produce a soft-set honey with a buttery texture. This is a technique you might like to try when you have a few years' experience. Most beekeepers sell only liquid honey which has been stored in bulk containers, has crystallised over time and has then been reliquefied before being jarred.

Warming honey

To liquify crystallised honey, most beekeepers use a honey warming cabinet. This is a thermostatically controlled heated box in which buckets of solid honey are warmed until the crystals have melted and the honey is liquid once more. It is important not to overheat honey or to heat it for too long, otherwise an organic compound called HMF (hydroxymethylfurfuraldehyde, for those who really want to know) can form, while an enzyme called diastase, which is created and added to honey by bees, will degrade. These can both be measured by your local trading standards department, and if they fall outside acceptable limits your honey will be declared unfit for sale. Excessive heating also impairs the delicate taste, aroma and colour of honey.

Honey heating cabinets are available to buy or you can make your own (see Further reading for details of plans). Liquifying a bucket of solid honey takes 24–48 hours at a temperature of 42°C (108°F). Be careful not to heat it any higher or for any longer than necessary.

Once re-liquified, honey can be poured into a settling tank to be jarred after 24 hours, or filtered through 100- or 200-mesh straining cloth (available from beekeeping suppliers) to remove any remaining crystals or particles

of dirt before jarring. Honey treated carefully in this way will retain its fine qualities and remain liquid for some months before crystallising in the jar.

Jars

Traditional honey jars hold one pound (16oz/454g), although UK law stipulates that they must be labelled and sold primarily in grams. Most beekeepers now sell their honey in ¾ pound (12oz/340g) jars. Honey for sale should be packed using new jars and lids. The jars should be washed in a dishwasher or in warm soapy water and rinsed and air-dried before use.

Selling honey

Honey is deemed a low-risk foodstuff and if sold in small quantities receives very little attention from local trading standards or environmental health departments. Providing you pay attention to standards of cleanliness and treat honey carefully and respectfully during harvesting and processing, you should have nothing to worry about when selling in small quantities to the public. In the UK, you should follow the requirements set out in the Honey (England) Regulations 2015, which can be found online. You do not need to register with your local authority as a food producer, although you can do so voluntarily. Though not mandatory, it is worth taking a short online course to obtain a Level 2 Food Hygiene Certificate, to acquaint yourself with general food handling, preparation and storage standards.

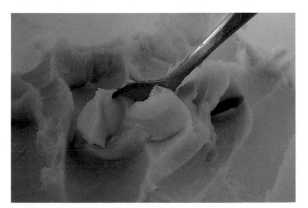

ABOVE Honey that has crystallised in the bucket. It will need warming before it can be jarred.

ABOVE Nicely labelled set honey.

If you do sell your honey, the main legal consideration is to comply with labelling regulations that stipulate the way in which your honey and its origins are described, as well as finer details such as the use of best before dates and the size of font used for certain parts of the label. Details can be found online, but you can now buy personalised legally compliant labels in a range of attractive designs from most of the beekeeping suppliers.

Beeswax

Though much emphasis is placed on honey, beeswax is also an important and valuable product of the hive. Bees invest a great deal of their time and resources in making wax, so it should be put to good use by the beekeeper and never wasted. Beeswax can be used for making candles, polishes, cosmetics, food wraps and more. There are three main sources of wax: burr comb, comb wax and cappings wax.

Burr comb is built in odd places around the hive, usually when the frames have been filled with comb and the bees want to continue expanding the nest. These small, misshapen sections of comb are found on the top and bottom bars of frames, underneath crownboards and in other nooks and crannies. Scrape them off at each inspection and collect in a lidded bucket

ABOVE Straining cappings wax.

ABOVE Odds and ends of collected wax.

BELOW A solar wax extractor.

– it is surprising how much burr comb can accumulate throughout the season.

Comb wax is reclaimed from super combs that are too old or damaged for re-use, or from ageing brood combs that have been replaced by frames of fresh foundation – it is good practice to replace a few brood combs each season so that none are ever used for more than three years.

There are various ways to melt the wax from old combs. One method is to use a solar wax extractor – a glass-lidded box that uses the power of the sun to melt and coarsely filter the wax. Another method uses steam generated by a DIY wallpaper stripper. Equipment for either method is commercially available or can be homemade. The wax collected in this way needs further filtering and refinement before it can be put to good use. Blocks of unrefined wax can be traded with beekeeping suppliers in return for new wax foundation or equipment.

ABOVE Unmoulding a wax candle.

Cappings wax, sliced from super frames during extraction, is the best-quality wax. It can be used to produce blocks of buttery yellow, fragrant wax that is ideal for making balms and beeswax wraps. After being removed from the comb, cappings can be left in a sieve for a day or two to drain – yielding yet more honey. They can then be soaked in water to remove any remaining honey. When drained, the resulting water and honey mix can be fermented to make mead. The clean cappings wax can be melted until liquid in a bain-marie on a stove, or in a Pyrex jug in a microwave using short bursts at medium power. Never melt wax directly over a naked flame – it's highly flammable. The molten wax is filtered through fine, lint-free material (bamboo nappy liners are excellent) into containers or moulds where it will set. Cakes of cleaned cappings wax can be used to make a range of products, or sold or swapped with fellow beekeepers who will make use of them.

Propolis

Used by bees to both strengthen and cleanse their nests, the healing properties of propolis have been familiar to people since ancient times and are increasingly recognised by modern medicine. A wide range of propolis-based products are commercially available, from toothpaste to skin ointments. The easiest product to make at home is a tincture, which has a wide range of uses and is particularly good for sore throats. Scrape excess propolis from frames, hive parts and queen excluders during hive inspections, or add a commercially available propolis screen, which the bees will gum up with harvestable propolis. Recipes and techniques can be found in specialist books and online.

ABOVE Scales of wax protruding from the wax pockets on a worker's abdomen.

Caring for bees in winter

Feeding bees

The main reason for feeding bees is to ensure they have sufficient stores to see them through the winter. Honey bees have the potential to be entirely self-sufficient, so feeding is usually only necessary if the beekeeper has harvested so much honey that there is little left for the bees to overwinter on. Most hobby beekeepers usually strive for a balance, taking only honey deemed surplus to the requirements of the bees and leaving enough so that feeding is unnecessary or kept to a minimum.

In most years, the honey stored in supers can be harvested in late spring and again in about mid-summer in the knowledge that the bees will be able to gather enough late-summer and autumn nectar to fill the brood box with sufficient reserves for winter. Late-flowering wild plants, especially ivy and Himalayan balsam, are particularly valuable in this respect. If it is cold or wet when these plants are in flower, though, feeding is likely to be necessary.

In late summer and early autumn, regularly assess how much honey is in the brood box. As the queen slows her egg-laying in the approach to winter, areas previously occupied by brood should become filled instead with honey – this is called backfilling. A full-sized colony needs at least 18kg (40lb) of honey to see it through the winter. The amount will vary depending on the exact size of the colony and the length and severity of the winter. You can assess stores by eye or by weight. As a guide, a British National deep brood frame holds about 2.25kg (5lb) of honey when full. Some beekeepers weigh their brood boxes with luggage scales, but most assess stores through a combination of examining frames by eye and by hefting – lifting a hive from the back with one hand to gauge how heavy it feels. A hive going into winter should feel almost as if nailed to its stand. As winter progresses, hives can be hefted every month, and in late winter every couple of weeks so to get an approximate idea of available stores. It can take a few years to perfect this technique, but it is quick and effective.

In early autumn (typically about mid-September), check the stores and if you think they might be insufficient for winter, feed the bees. Syrup fed to bees should be as heavy (thick) as possible so that they don't have to spend time and energy processing it before storage. Syrup should only be fed when temperatures are warm and the bees are still active enough to remove excess water and store it like honey – as a guide, when daytime temperatures are still regularly above 10°C. Ready-made syrup can be purchased, or you can make your own. To make about two litres of heavy syrup, tip 2kg of white table sugar (never brown sugar or honey) into a mixing bowl, then add 1.2 litres of boiling water from the kettle. You don't have to be too precise about quantities. Stir until the sugar is dissolved, then pour into buckets or bottles for transfer to the hive. Honey produced and stored in comb by the bees themselves is the best food, but stores topped up with syrup are perfectly adequate.

LEFT Although few signs of life are seen on the outside, in winter a colony will continue feeding, generating warmth and even rearing a few young.

RIGHT Hefting a hive.

ABOVE **Making syrup.**

ABOVE **Topping up a small rapid feeder.**

ABOVE A block of fondant placed over the feed hole.

ABOVE Fondant bag emptied and licked clean.

Feeders

A variety of feeders are available for use in different situations. For the fast feeding of large quantities of syrup during the approach to winter, rapid feeders are best. These are placed over the crownboard above the brood box and are contained within an empty super. Bees move up into the feeder through the feed holes in the crownboard and climb over a ridge to reach the syrup. Circular rapid feeders take about 2 litres of syrup and are fine if the hive is close to home and can be refilled regularly. The larger English, Ashforth or Miller feeders take up to 10 litres of syrup and do not need to be refilled as frequently – ideal if your hives are some distance from home.

Add syrup to feeders at dusk. Trickle a little syrup down the access hole of the feeder into the hive to make a trail that the bees will follow up into the feeder, but be careful not to spill any syrup on or around the hive. If you do, clear it up or wash it away. This will help to prevent robbing, where your own bees or those from elsewhere will begin searching for the source of the syrup and might try to force their way into hives with feeders on them, causing a dramatic war between bees.

If the colonies that have been fed want the syrup, it will be taken down into the hive remarkably quickly. Check again the following evening to see if more is needed. Keep giving syrup until the bees no longer take it, at which time they may have filled all of the available space in the brood box. Aim to have all feeding done within a week or so, and before temperatures fall to the point where bees are no longer regularly active, usually in late autumn.

Other reasons to feed

In a long winter, or a mild one when bees fly frequently and consume more energy, colonies might run out of stores. It is particularly important to monitor stores in late winter and early spring when the queen begins laying eggs and the colony starts to expand, but little nectar is available for collection from flowers. If you are worried about stores being insufficient, you can feed colonies with fondant – a sugar paste that bees can consume as needed. Fondant (also called candy) is sold in polythene-wrapped blocks. Cut a hole in the polythene and place the block over the feed hole in the crownboard or on the top bars of the brood box. The hole in the polythene should be directly over the bees. If the bees are hungry, they will come up and eat the fondant, licking the polythene clean, in which case it may need to be replaced. Some beekeepers feed fondant instead of syrup in late autumn, the bees storing the white paste in cells for winter use in the same way as they would honey or syrup.

Colonies with perilously low or no stores during the active season can be fed to keep them alive until flowers begin to yield nectar. This is called emergency feeding. A colony with no stores and weakened bees can be given immediate help by using a handheld spray bottle to mist bees on the comb with a light sugar syrup. The bees will lick this off each other for an instant energy boost. They might then be able to find sugar syrup supplied in either a frame feeder (a plastic container of syrup inserted in the brood box in place of a frame),

or a contact feeder (a sealed bucket with small holes in the lid which is placed upside down immediately over the cluster of bees). Both types are available from equipment suppliers.

Any small or weak colony might need a boost of sugar syrup at any time during the year. This could include a colony that has swarmed or a captured and housed swarm, a divided or nucleus colony or a colony being prepared by the beekeeper to raise new queens.

Preparing for winter

The approach of winter is a busy time for bees and beekeepers. All are concerned with preparing for the season ahead, giving colonies the best chance of survival and of entering the following year in a healthy condition. Tasks done in autumn are often thought of as the final jobs of the beekeeping year, but actually this is when the foundations of future success are laid. Rather than thinking of autumn as the end of the beekeeping year, it is perhaps better considered the start of a new one.

Feeding

By ensuring that hives have sufficient stores as described in the previous pages, the beekeeper gives colonies a good chance of surviving through winter until the first significant sources of pollen and nectar become available in early spring. It is possible to give bees emergency food supplies if needed, but this should not be necessary if bees can store plenty of food in autumn.

Insulation

Free-living honey bee colonies thrive in tree hollows with very thick timber walls. With their nest so well insulated, it is remarkable how little food such colonies consume in order to keep warm over winter. Most commercially available wooden hives have relatively thin walls and are much less thermally efficient. Bees can overwinter adequately in such hives, but consume more stores. Polystyrene hives are more thermally efficient. If you have wooden hives, a simple form of additional insulation is a sheet of foil-backed expanded foam insulation material placed in a super above the crownboard. Alternative materials include sheep's wool or wood shavings. Insulation above the crownboard

ABOVE Insulating foam over a brood box.

prevents a great deal of the bees' expensively generated heat from escaping. It also keeps the crownboard from getting cold, which can cause condensation to form on the underside and water to drip down into the nest. Bees generally cope well with cold but not with damp.

Pests and diseases

In late summer and early autumn, bees are often treated for the parasitic mite varroa. Treatment at this time means that varroa will not affect the production of healthy bees that are needed to survive the winter. It is also when precautions can be taken to protect colonies against damage from wasps, mice and woodpeckers. This is also the time to be particularly vigilant for the invasive Asian hornet. These threats, and measures to protect against them, are covered in the following pages.

Strong colonies

If you have any small colonies that you fear won't make it through winter, colonies with old or failing queens, or just too many colonies as a result of swarming, consider uniting some of them (see page 103). Make sure that only healthy stocks are united. After uniting, reorganise frames of brood and stores from the two halves so that the best ones remain in the new, stronger colony.

BELOW Strap down hives on exposed sites.

Other winter preparations

Check the woodwork of hives to make sure they are secure and weatherproof. Ensure that hive stands remain strong and are on a firm footing. Cut back vegetation that has grown around hives. Trim any overhanging branches that might fall in a storm. If hives are in an exposed area, consider strapping the hive components together. A ratchet strap around the hive and stand will help to keep everything together in high winds. In very exposed areas or when severe winds are forecast, hives can be strapped to the ground using ties that are screwed down into the soil. At the very least, hives with shallow roofs and nucs can be made more secure with a brick or large stone to hold the roof down.

Bees in winter

The bees will have made their own preparations for winter. Using nectar from late-flowering plants and any syrup provided by the beekeeper, they will have worked to fill brood combs with capped stores. Energy-giving carbohydrates in the form of honey or processed syrup are their main requirement over winter. They will also store modest quantities of pollen to provide protein for the small amount of brood produced in winter and early spring.

Propolis will have been used by the bees to seal any cracks and crevices, possibly even partially closing the hive entrance with a propolis curtain. They are battening down the hatches.

Although the queen's egg-laying tails off after its peak in early summer, there is usually a secondary peak in late summer and early autumn. The young workers produced at this time are subtly different in that they carry more fat and remain youthful for much longer. These are winter bees, which will live for perhaps six months rather than the usual six weeks. They will help to keep the colony warm, and feed and care for brood when egg-laying commences in earnest again with the approach of spring.

Honey bees do not hibernate, but as winter sets in and temperatures drop, they will begin to cluster, huddling together to keep warm and protect the queen and any remaining brood. The bees consume the stores in the comb where they are clustered and vibrate their wing muscles (in other words, shiver) to produce heat. If brood is present, the core temperature of the cluster

ABOVE The winter cluster.

is kept at about 34.5°C. Without brood it is maintained at about 21°C. The energy used to keep the nest warm throughout the winter is said to be equivalent to continuously running a 20-watt lightbulb, showing why it is essential for the bees to have such a well-stocked store cupboard. The size of the cluster will change with temperature. The colder it becomes, the tighter the bees huddle, spreading out again in warmer conditions. The cluster will gradually change position, moving through the brood box as the stores in each area are consumed.

On warmer days, bees will fly from the hive, sometimes to find scarce sources of winter nectar and pollen, but more often to empty their bowels, which can balloon to hold the accumulated waste of weeks or months. If thousands of your bees take cleansing flights on the day your neighbour hangs out their bedsheets, it could cause a diplomatic incident; bee poo is colourful, very sticky and can sometimes fall in significant quantities.

Workers also collect water, essential to dilute honey before it can be consumed. If you haven't done so already, establish a reliable source of shallow water near the hives, ideally where it might catch some sun and warm a little. If you get the chance, visit your hives on warmer winter days. It is a joy to watch the bees flying and perhaps bringing in a little pollen after weeks or months without seeing them.

After usually having had a break in mid-winter, the queen starts to lay eggs again when daylight hours begin to lengthen. Workers keen to collect fresh pollen to feed the young will seek out early flowers, including hellebores, snowdrops and crocuses. The first significant source of pollen is usually hazel, but the later-flowering pussy willow (goat willow – *Salix caprea*) can give colonies a significant boost with plenty of golden pollen and some nectar too. As better weather and more spring flowers bring increasing opportunities to gather nectar and pollen, colonies begin to expand, ready for the season ahead.

ABOVE Even in freezing conditions, healthy bees will be snug and warm inside their hive.

Health and hygiene

As with all livestock, bees can be affected by various pests, parasites and diseases. The idea of this can be off-putting to new beekeepers, but most health issues are little more than a nuisance and can easily be addressed. A few are very serious, but thankfully these are rare. Beekeepers must know how to recognise such threats to their colonies and deal with them appropriately.

Basic apiary hygiene

There are few veterinary treatments for bee diseases, but most problems can be prevented, minimised and sometimes solved by good beekeeping management and biosecurity practices.

Pollen, honey, wax and hive parts may contain disease, so the transfer of these between apiaries should be avoided. Most beekeepers treat each apiary as a separate, self-contained unit and avoid the swapping of bees and equipment to minimise the transfer of any disease. Brood combs in particular can act as reservoirs for disease, which is why partial annual replacement is recommended or periodic complete replacement using either a shook swarm (see page 153) or a Bailey comb change (see page 156).

If you see signs of any kind of infection, thoroughly wash your tools and change your gloves and bee suit before inspecting further colonies. When it comes to cleanliness, washing soda crystals (sodium carbonate) are the beekeeper's best friend. They can be used to clean equipment and bee suits, and are about the only thing that will dissolve propolis. Keep a lidded bucket containing a strong solution of washing soda (1kg to 5 litres of water) with an added squirt of washing-up liquid, and take it with you when you inspect hives. Hive tools should be washed and scrubbed with a scouring pad between colonies. If you are wearing latex gloves, these can be cleaned between colonies while still being worn. Bee suits should be washed frequently in beekeeping season. Add a cupful of washing soda crystals to the washing machine along with your usual detergent when doing so. Washing soda crystals are a natural product and can safely be washed down drains.

It is worth having access to a reserve apiary site at least 3 miles (5km) away that can be used for quarantine. If you collect a swarm of unknown origin or buy new bees, keep them in quarantine until you have had a chance to inspect sealed brood for disease, before introducing them to your home apiary.

Apiary design can help to prevent disease from being spread by drifting (bees accidentally entering the wrong hive) or robbing (bees stealing honey from another colony). Hives are best arranged so that they are at least 1m (3 feet) apart, are not in a straight line and have variously oriented entrances.

The National Bee Unit

In England and Wales, the government-funded National Bee Unit (NBU) is responsible for overseeing bee health and running various bee health programmes. They employ a team of local bee inspectors who visit apiaries to identify and deal with serious bee diseases. BeeBase, the NBU website, has excellent in-depth advice and information on bee health. Beekeepers who register their apiaries receive topical information about disease outbreaks and starvation in their area. Other countries have their own equivalent of the NBU.

Brood diseases

The majority of honey bee diseases affect brood and are found by examining brood comb during regular colony inspections.

LEFT Guard bees defending a reduced hive entrance against unwelcome intruders.

SIGNS OF A HEALTHY BROOD

Eggs Honey bee eggs resemble tiny grains of rice. They are white, slightly curved and can stand upright or at an angle. A single egg is laid at the centre of the base of each cell. Where there is one cell containing an egg there should usually be many more in close proximity. Good queens in a healthy colony will lay in a consistent, solid pattern with few gaps. A spotty brood pattern should prompt you to look closer at what is going on.

Larvae Newly emerged larvae are tiny, curved and semi-transparent. They lie in a small puddle of whitish brood food. As they grow, they turn a glossy, creamy-white colour and the segmentation of their bodies is clearly visible. Larvae should always lie flat on the bottom of the cell in a 'C' shape, quickly growing to fill the cell nose-to-tail before moving into the vertical position for pupation.

Capped cells During pupation, brood cells are capped with a mixture of wax (usually recycled), pollen, propolis and possibly other materials scavenged by the bees. The cappings are porous to allow air and pheromones to pass through them. They can range in colour from sandy yellow to dark brown, but should always be convex (domed) and dry-looking.

ABOVE Eggs.

ABOVE Capped cells.

ABOVE Larvae.

Minor brood problems

The following relatively minor problems, not all them actually diseases, can be found by observing the brood.

Chalkbrood

A fungus, *Ascosphaera apis*, kills larvae after the cell has been sealed. The bees uncap the cells to remove the dead larva, so you might see white, flattened-looking 'mummies' in cells. These are removed and thrown out of the hive by the bees, sometimes accumulating on the hive floor and looking like pellets of white or grey chalk. Fungus grows on the mummies, making them turn grey or black. You might notice a spotty brood pattern where dead larvae have already been removed.

Finding a few chalkbrood-infected cells in spring is not uncommon, and often the infection seems to clear up as the colony expands. Well-fed, unstressed colonies withstand chalkbrood infection better. Re-queening is helpful where persistent chalkbrood infections occur as some honey bee strains are more resistant than others. General apiary hygiene as outlined earlier will help to prevent recurrence and spread.

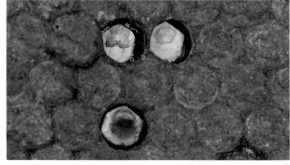

ABOVE Chalkbrood mummies still in their cells.

Sacbrood

A virus prevents pupating larvae from moulting their skin for the final time when going through metamorphosis. Trapped in their own skin, they die and disintegrate, producing a fluid-filled sac. The bees will uncap the infected cell and you may see the pointed end of the sac poking upwards. The bees usually remove the sacs, leaving empty cells. Sometimes the sacs dry in the cell, leaving a brown scale which can be removed easily.

As with chalkbrood, it is not uncommon to see a few cells of sacbrood, particularly in spring, but infections usually clear up as colonies expand. Severe cases can be addressed by re-queening.

Bald brood

Sometimes you will find uncapped cells containing pupating larvae, with the semi-formed bee clearly visible inside. Some strains of honey bee uncap brood to check if varroa is present, recapping after inspection. This is now considered a good behavioral trait, and can result in a scattered pattern of uncapped or semi-capped cells. Another cause can be caterpillar-like wax moth larvae, which eat the cappings, burrowing their way across a frame and munching the wax as they go. They sometimes leave a papery-looking white skin over each cell and other times uncap them completely. The larvae usually travel in a straight line, so you may see several affected cells in a row. Some colonies fail to cap their brood from the outset, leaving occasional cells or sometimes even very large patches of pupating larvae uncapped. This is usually a genetic problem, possibly caused by inbreeding, and can be addressed by re-queening. See page 143 for how to deal with varroa and wax moth.

Chilled brood

Bees maintain the temperature of their brood nest at around 34.5°C. If temperatures fall significantly below this, the brood will die. Sometimes, particularly in early spring, the queen can lay more eggs than the workers are able to adequately care for. In a sudden cold snap, the workers will cluster over as many eggs and larvae as they can. Brood that is not covered will chill and die. Dead eggs and larvae will usually be removed but can sometimes be left for a while, shrivelling and turning brown or black. The death of some brood is usually only a temporary setback, and colonies soon recover. Having well-insulated hives can help to reduce losses.

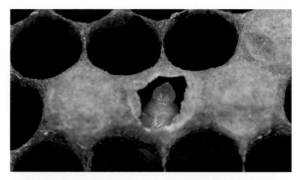

ABOVE Uncapped sacbrood.

ABOVE Sacbrood removed with tweezers.

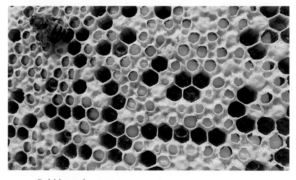

ABOVE Bald brood.

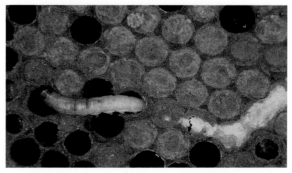

ABOVE Wax moth larva in the process of uncapping brood.

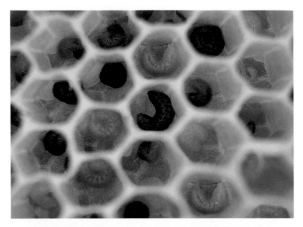

ABOVE Chilled brood, not uncommon in early spring.

ABOVE Drones emerging from scattered, dome-capped cells – the result of a drone-laying queen.

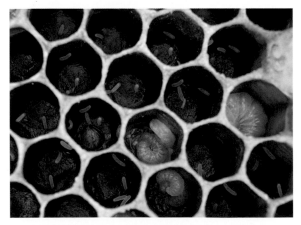

ABOVE Multiple eggs and larvae per cell – the signs of laying workers.

Drone-laying queen

This is not a disease, but an alarming sight and something that needs to be addressed immediately. Rather than capped cells being slightly convex, they rise up in exaggerated clusters that look like a miniature mountain range. This is usually the sign of a drone-laying queen – that is, a queen who has run out of sperm due to old age or inadequate mating. She will continue laying eggs, but they will not be fertilised and will produce only drones. The solution is to re-queen the colony.

Laying workers

Although workers do not normally lay eggs, they do have ovaries. These are usually latent – their growth and functionality suppressed by pheromones produced by the queen and by open brood. After about a month of having no queen or open brood in a colony, the ovaries in some workers become active and they begin to lay eggs. Laying workers usually deposit several eggs in each cell in an untidy manner, often both on the side walls and the base of the cell. If the eggs hatch and larvae develop, the comb will have a scattering of cells containing multiple larvae. These will all be male because the egg-laying workers will never have mated. Pupating drone larvae will cause raised cell cappings similar those seen with a drone-laying queen, though usually more scattered.

It is difficult to solve the problem of laying workers, and it is best to try to prevent it from happening in the first place. It is not uncommon for queens to go off lay for a time in the summer, so if you don't see any eggs and cannot find the queen, this does not necessarily mean the queen has died. If you are unsure, place a brood frame containing eggs and young larvae (no bees) taken from another healthy colony into the hive. If no queen is present, the open brood will help to suppress the development of workers' ovaries, and they should use the material you have given them to make emergency queen cells and produce a new queen.

If laying workers are well-established, they will not try to make a queen from any brood you give them and are unlikely to accept a new queen. The only realistic solution is to move the hive to one side and shake all of the bees off the frames. Then take the empty hive away. The bees will fly into the nearest hive and will boost the numbers there. It is a sad end to a colony, but it is not uncommon and shows why it is always worth keeping more than one colony.

Serious brood diseases

Two diseases that affect honey bee brood are very serious indeed. They are rare, but it is important that you can recognise the symptoms. In the UK, both of these diseases are notifiable *by law* – you must immediately inform your local bee inspector if you suspect your bees have either. If concerned, the inspector will visit within a few days to check and possibly test your bees. If you receive a positive identification, you will be given instructions on the best course of action.

European foulbrood (EFB)

This is the more common and easily spread of the two major brood diseases. European foulbrood (EFB) is caused by the bacterium *Melissococcus plutonius*, which infects the gut of larvae before they pupate. Larvae usually die and rot in their cell, and this is the most obvious sign. Some infected larvae may survive and produce adult bees. EFB is most likely to be found in spring when there are fewer bees in the hive.

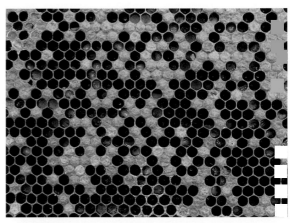

ABOVE A pepperpot brood pattern suggests a problem; in this case EFB.

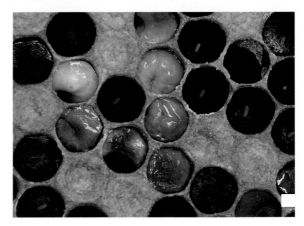

ABOVE Twisted, discoloured, decomposing larvae are a symptom of EFB.

SYMPTOMS

- Dead or dying larvae will lie in awkward-looking positions in their cells rather than in the usual 'C' shape.
- Dead larvae appear melted and pulpy, losing their clearly defined segmentation.
- Dead and dying larvae range from a creamy colour to yellow and dark brown.
- Sometimes there will be only one or two instances in an entire frame of comb, but at other times there will be many.
- The bees may remove infected larvae, leaving a patchy brood pattern.
- Larvae left in the cell may dry to form a scale that is easily removed from the base of the cell.
- There may be an unpleasant 'rotting' smell in the hive.
- Larvae sometimes die after capping, in which case the cappings will appear sunken and may be perforated.

Some of the above symptoms can have other causes, such as a severe infestation of varroa (see page 143) but if you are worried, contact your local bee inspector. A lateral flow test can confirm infection within minutes. It may be recommended that the colony be destroyed. If the colony is large and it is early enough in the season, it might be recommended that you perform a shook swarm (see page 153).

EFB spreads easily and if one colony in an apiary has it, then it will probably be present in others too. It can be transmitted when bees rob honey from weakened, infected colonies, and can be carried by swarms. Beekeepers are probably a major cause of spread, swapping infected equipment and using contaminated hive tools, gloves and bee suits.

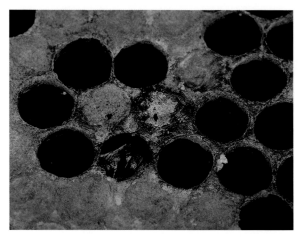

ABOVE Sunken, perforated cappings are a symptom of AFB.

ABOVE Ropey remains suggest AFB. Burn the matchstick in your smoker.

American foulbrood (AFB)

American foulbrood (AFB) is rarer than EFB. The spore-forming bacterium *Paenibacillus larvae* infects the gut of developing larvae and kills them after the cell has been capped. Millions of spores are produced inside the rotting pupa, which cleaner bees ingest and then pass on in brood food. The spores can lie dormant for decades, surviving disinfectants or extremes of temperature, which is why it is inadvisable to buy second-hand equipment, however clean it looks. A major source of AFB is imported supermarket honey, particularly from China (though you won't find Product of China on any labels). Honey bees fly in through kitchen windows or find unwashed jars put out for recycling, ingest infected honey and take it back to their colony.

ABOVE Bee inspector using a lateral flow test to diagnose foulbrood.

SYMPTOMS

- Sunken, perforated or ragged looking brood cappings that may appear wet or greasy.

- A bad smell, though not always present.

- A 'pepperpot' brood pattern where cells have been uncapped and remains removed.

- Empty cells may contain a dark scale at their base which cannot easily be removed.

- You might also see the dried remains of the dead pupa's tongue.

A simple initial test for AFB is to poke a matchstick into an infected cell, stir up the larval remains and then slowly draw it out. If the remains form a caramel-like rope of up to 3cm (1 inch) long, it is likely that you have a positive case. Lateral flow tests are available for more scientific testing, but if you have performed a positive rope test your next move should be to contact your bee inspector and let them guide you. In the meantime, reduce the hive entrance to one bee space (less than one finger's width) to lessen the chances of bees from other colonies entering the hive and robbing infected honey from the weakened colony. Do not inspect any other colonies until you have a clean hive tool, bee suit and gloves.

In the UK there is no legal medicinal treatment for AFB. Colonies must be destroyed and their remains burned and buried. Bee inspectors will oversee this and the disinfection or destruction of hives and equipment.

Brood inspections

It can be difficult to see evidence of brood diseases because brood combs are usually covered with bees. With EFB in particular there might be just one or two infected cells on a frame. Regular inspections are an opportunity to look for disease, but it is wise to conduct a thorough disease inspection at least twice each season – once in early summer and again in early autumn. To do this, remove the outermost frames of stores from a brood box, brushing or shaking any bees back into the box. When you reach a frame of brood, hold the lugs very firmly in each hand and position the frame so that it is over and partially within the space you have created in the brood box. Then give it a very sharp downward shake without knocking it into the hive or any other frames. The majority of the bees will fall into a pile on the floor of the hive and will not be harmed. With most of the bees removed from the frame, check every brood cell. Use a bright light or make sure the sun is behind you so that you can see to the bottom of cells. A magnifying glass can help. When checked, replace the frame so that the bees can repopulate it. Then move on to the next one. Keep an eye out for the queen when doing this. She is unlikely to be harmed by being shaken off the frame, but for added safety she can be caged and reintroduced when the job is complete.

Adult bee diseases

Some diseases affect adult bees and can be identified through the appearance or behaviour of the bees themselves or of the colony as a whole. The most common and disruptive are as follows.

Nosema

This is one of the most significant bee diseases, and it probably affects colonies more often than many beekeepers realise. A parasitic microsporidian pathogen called *Nosema apis* invades digestive cells in the gut of the honey bee. Spores are then excreted and picked up by other bees, spreading the infection.

Nosema affects the ability of bees to digest pollen, making adult bees weaker and shortening their lives. It also reduces the production of brood food and therefore the feeding of larvae and the queen. *N. ceranae* is a more

SYMPTOMS

- Although appearing to be largely normal, colonies fail to build up significantly in spring. Sometimes they develop only slowly, or dwindle away altogether.

- Although not a definitive symptom, colonies may suffer dysentery as a secondary infection. Spots or streaks of faeces may be seen inside the hive, and on its outer hive walls and roof.

ABOVE A sample of 30 bees is crushed before testing for nosema under a high-power microscope.

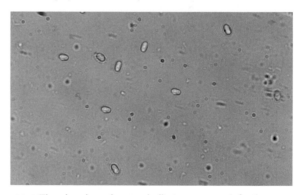

ABOVE Tiny rice-shaped spores indicate a nosema infection.

virulent strain that developed in the Asian honey bee and now also affects the western honey bee.

Positive identification can only be achieved using a microscope. Most beekeeping associations will have someone who can study a sample of your bees. There is no licensed medical treatment in the UK. Changing the brood comb and hive parts with a Bailey comb change (see page 156) can remove spores. Stress factors such

ABOVE Yellow faeces outside the hive might suggest nosema. Faeces inside the hive, however, makes it much more likely.

as damp apiary sites, lack of nutrients, crowded hives or infection with any other disease can contribute to nosema outbreaks.

Chronic bee paralysis virus (CBPV)

An increasingly problematic virus, CBPV causes two forms of symptoms in adult bees. With the most common form, affected bees show an abnormal trembling of the body and wings, have bloated abdomens and dislocated or distorted wings, and they crawl on the ground outside the hive or on the top bars inside the hive. The other form is identified by the presence of hairless, shiny, black-looking bees, which are attacked and rejected from the hive. Both forms of symptoms can be seen in bees from the same colony. In many cases the condition is noticeable in spring and early summer but disappears through the course of the season. If not too badly

affected, colonies often recover. In serious cases, CBPV can cause the rapid death of many bees, which may be found in piles outside the hive entrance or on the hive floor. This can be confused with poisoning.

There is no medical treatment for CBPV. In crowded conditions it will spread more rapidly because bees will rub up against one another, pulling out hairs from their abdomens and leaving open wounds through which the virus can enter. The virus therefore worsens after bad weather in spring and summer, when hives are highly populated and the bees are confined. Adding an extra brood box or supers to provide space can help. Although time-consuming, individually picking out symptomatic bees and disposing of them can help to remove the cause of further infection, as can temporarily removing the hive floor so that dead bees drop to the ground and can be swept up and removed.

Other problems

Acarine is a condition caused by a tiny mite, *Acarapis woodii*, that infests the tracheal tubes of the bee. Once problematic, it is now rare.

Amoeba is caused by an organism called *Malpighamoeba mellificae*, which affects the malpighian tubes, the bee's equivalent of kidneys. Amoeba often occurs at the same time as nosema and can be treated in the same way.

Colony Collapse Disorder (CCD) is not a disease but a description of the way many colonies have died suddenly from unknown causes. The phrase was coined in the early 2000s in the US, where the most dramatic instances occurred. The exact causes remain unknown, but it is thought likely to involve a combination of

ABOVE A dark, shiny and hairless victim of CBPV.

ABOVE Dead bees with their tongues poking out suggests poisoning.

factors, possibly including nosema, varroa and related viruses, poor nutrition, and the stressful management techniques often practised by large, commercial beekeeping concerns. Other theories have blamed neonicotinoid pesticides and interference from mobile phone masts. CCD is not a significant concern for small-scale hobby beekeepers.

Poisoning

Chemicals used on farms and in gardens can have significant detrimental effects on bees (see pages 14–15). Bees foraging on a crop that is sprayed with harmful chemicals may return to the hive to die and will be found dead on the comb, on the floor or in a heap outside the entrance. One sign of probable poisoning is dead bees outside the hive with their tongues sticking out. Report suspected poisoning to your local bee inspector, who may initiate an investigation.

Varroa

The parasitic mite *Varroa destructor* is undoubtedly the greatest cause of colony ill health. Varroa originated in Asia as a parasite of the Eastern honey bee, *Apis cerana*.

It jumped species to the imported western honey bee, causing far greater problems for a bee that had not evolved any defences. Varroa then spread from east to west as colonies were moved by humans. It was first found in the UK in 1992, and varroa is now endemic in most parts of the world. If you have honey bees, they will almost certainly have varroa – even if you can't see the mites. Since 2021 it has been a legal requirement to report the presence of varroa in an apiary to the National Bee Unit.

Varroa affects both brood and adult bees, and can cause the total collapse of colonies if not addressed by the beekeeper. There is some division between the majority of beekeepers who treat their bees for varroa and a small but growing number who don't. Non-treaters believe that honey bees have – or are capable of developing – their own defences against varroa, and that if allowed to continue doing so will evolve immunity or coping mechanisms. There is growing evidence that some colonies can cope with varroa, and research is ongoing. For most beekeepers, though, it remains a fact that varroa-infested colonies left untreated are likely to collapse and die.

Varroa causes problems on multiple levels for bee colonies. Adult varroa mites, which look like tiny

ABOVE Varroa on bees.

reddish-brown crabs, live on the bodies of bees. They use their sharp mouthparts to pierce the bee's softer parts and feed from them, consuming their body fats and haemolymph (blood) and weakening them. Immature varroa live in brood cells where they feed on the developing bees, transferring viruses to their host. A range of viruses are passed on in this way and it is these that weaken and kill colonies. Colonies with heavy infestations of varroa can have multiple symptoms, with both adult bees and brood suffering in various ways – a condition known as parasitic mite syndrome (PMS).

Varroa life cycle

Varroa mites reproduce among honey bee brood. Adult females enter brood cells shortly before bee larvae pupate, and are sealed inside when the cells are capped. There they lay eggs that hatch into tiny white nymphs which grow and mature while feeding on the developing bee larva. Sometimes this process kills the young bee. Sometimes they survive but after pupation emerge weakened, damaged or infected. When one mite enters a cell, two or three, including the mother, will emerge about two weeks later. Slightly more will emerge from drone cells than from worker cells, because they are sealed for three days longer. Varroa mites therefore target drone larvae. This may not seem like a high reproduction rate, but it means that a dozen or so mites can multiply to thousands over the course of a summer.

Symptoms

It is not always easy to find varroa in a colony, even though at about 1.6mm (1/20in) across they can be seen with the naked eye. Mites hide under the overlapping

ABOVE Deformed wing virus.

ABOVE A dead, newly emerged worker. A varroa mite is poking out from beneath an abdominal segment.

abdominal segments of adult bees, or in the hairy area between the abdomen and thorax. If you start seeing adult bees with varroa on them, you can be sure there is a significant infestation, especially as in summer about 70 per cent of the varroa in a colony are likely to be sealed within brood cells. The most likely visual symptom in adult bees is the presence of shrivelled, stunted wings – a condition called deformed wing virus (DWV).

A serious infestation will show its effects in the brood nest, where you may see some or all of the following:

SYMPTOMS

- Spotty, pepperpot brood pattern
- Nibbled or perforated brood cell cappings
- Chewed, headless pupae in open cells
- Dead pupae
- Dead emerging bees. Heads poking out of cells, often with their tongues extended
- Dead, melted-looking larvae and sac brood
- Supersedure queen cells.

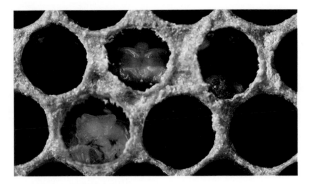

ABOVE Varroa-affected pupae.

You will notice a decline in bee numbers and a more spotty brood pattern as the colony heads towards collapse; bees may also become more aggressive during inspections.

Monitoring for varroa

There are various ways to monitor colonies and assess their varroa population. This should be done several times a season. One way is to estimate how many varroa are inside the cells of developing drones, since this is where varroa prefer to reproduce.

Drone brood monitoring

In mid-spring to early summer, when drones are being produced in large numbers, identify an area of capped drone brood where the drones are close to maturing – they should be white with pink eyes. Use an uncapping fork to pierce the cappings and lever out drone brood. You will need to study about 100 drone pupae. Sadly, they will have to be sacrificed. Study the pupae and count how many have varroa on them. An acceptable level of infestation is one varroa mite per 100 drone

ABOVE Forked-out drone larvae with varroa.

pupae. Three mites or more per 100 drone pupae suggests a serious infestation. The danger thresholds of this technique and those described below vary slightly throughout the season, so it is worth referring to the BeeBase website (www.nationalbeeunit.com).

Inspection tray monitoring

Most modern hives come with an open-mesh floor. This is designed to help to reduce varroa loads and also to monitor them, as follows:

- Coat the inspection tray with vegetable oil or Vaseline to make it sticky. Slide the tray under the floor and leave for at least seven days.

- Count the number of mites that have fallen on the tray (a magnifying glass is helpful) and divide the number of mites by the number of days since the tray was inserted to get an average daily mite drop.

- In summer, a daily drop of less than 10 mites is considered acceptable for the time being. Between 10–30 suggests that treatment will be necessary soon. A drop of more than 30 means immediate treatment should be considered.

ABOVE Checking inspection tray for varroa.

ABOVE Pollen, wax and varroa on an inspection tray.

Sugar-shake method

Probably the easiest method uses icing sugar to knock varroa mites off the bodies of the bees, making them easy to count, as follows:

- Use any container that has a mesh or perforated lid that will allow icing sugar but not bees to pass through. Beekeeping suppliers sell suitable containers with marks to measure bee quantities.

- Collect 300 bees (avoiding the queen).

- Add about 7g of fine, dry, icing sugar.

- Roll the jar gently so that the sugar coats the bees.

- Let the jar stand for one minute.

- Invert the jar and shake it over a white surface until all the icing sugar and no more mites fall through. This is best done indoors if breezy.

- Count the mites and work out the percentage.

- Return the bees to the hive.

ABOVE Sugar-shake test.

Controlling varroa

Most beekeepers use a system called Integrated Pest Management (IPM) to control varroa. The idea is that, although you are unlikely ever to totally eradicate varroa mites, using various methods to knock back their numbers throughout the year can keep them at levels that don't threaten colony collapse. There is quite an array of varroa treatment products on the market and choosing one can be somewhat daunting. This is one area where you will definitely need to do more research and keep up to date. Find out what trusted, experienced local beekeepers do, and consult beekeeping suppliers about what treatments they sell and how they should be used. The following is a typical IPM strategy.

Integrated pest management

Monitor colonies throughout the year for signs of varroa infestation. If moderate numbers of varroa mites are found during the summer, consider a biotechnical management method such as one of those below. These methods work best between the middle of spring and mid-summer.

Sacrificial drone comb

Add a frame or two of drone foundation to the brood box. Foundation that has embossed cells the correct size for drone comb will be drawn out into drone cells, and the queen will lay unfertilised, drone eggs in them. Varroa mites will be attracted to these cells. When the cells are sealed, remove the frames and destroy the comb. Alternatively, add a super frame of ordinary foundation to the brood box. The bees will build 'wild' comb down from the bottom of the shorter frame – it

ABOVE Removing drone comb from the bottom of a brood frame.

is likely to be drone comb, which can be cut off and destroyed when the cells are sealed.

Queen trapping

The queen is confined to a cage in the brood box that holds her and a single brood comb, meaning this is the only place she can lay. After nine days the comb is removed from the cage and placed in the main brood box, and the queen, still in the cage, is given another comb to lay in. This is done three times, every nine days. Varroa mites that want to reproduce will have no choice but to use only the three combs that the queen has had access to. As each of the combs becomes fully capped in turn, remove the comb and destroy it. When the third caged comb has been laid up, the queen can be released back into the colony to lay where she wishes.

Modified shook swarm

Perform a shook swarm (see page 153). Destroy all of the brood combs taken from the colony and replace with foundation except for two or three combs of empty, drawn comb. The queen will immediately lay in these combs while more comb is being constructed elsewhere. Varroa mites wishing to reproduce will invade these combs. When the brood combs are sealed, remove and destroy them. Replace with frames of foundation.

Late-summer treatment

Knocking varroa back with the above methods is effective, but most beekeepers supplement this with a twice-yearly dose of an organic acid or essential oil-based treatment, possibly alternating them with synthetic 'hard' chemicals. This is typically done as follows.

ABOVE Thymol-based varroa treatment.

Most treatments cannot be administered when honey destined for human consumption is in the hive, so must be used only after the honey harvest. Treating as soon as the honey has been harvested in late summer means that the colony will have low numbers of varroa mites when healthy winter bees need to be produced. So-called 'soft' treatments contain various organic chemicals such as thymol, which is based on an extract of thyme. Most soft treatments are approved as organic by the Soil Association (although you still can't label your honey as 'organic'). 'Hard' treatments contain synthetic varroacides, such as tau-fluvalinate or amitraz.

Each treatment type has a slightly different treatment regime. Typically, a dosing strip or container is placed in the brood box, followed by a second dose two weeks later, taking four weeks in total. The idea is that the active ingredient is present in the hive for longer than one brood cycle, so that every varroa mite in the colony will at some stage be exposed to it; varroa mites sealed inside cells are not usually exposed to the chemicals, which cannot penetrate the wax cappings. Make sure you follow the instructions precisely and remove the dosing device at the end of the period. Using more than the suggested dose or for longer will not make it more effective.

In the UK there are strict rules about using only treatments approved by the Veterinary Medicines Directorate (VMD). If you buy treatments directly from reputable beekeeping supplies companies, you will not go wrong. Because you are using veterinary treatments on an animal used to produce a foodstuff – UK law requires you to record and retain details such as batch numbers and dates used. The BeeBase website has a downloadable treatment records sheet.

Winter treatment

Late-summer treatments can knock back a high percentage of mites in a colony, but a further treatment in winter will give additional protection. Winter treatments use oxalic acid, an organic compound found naturally in plants like rhubarb and spinach. Again, it is approved for use by the Soil Association. Oxalic acid treatments are used to treat varroa living on the bodies of bees, and will not reach any that are reproducing in sealed brood cells. Therefore, they should be used in mid-winter when there are few – ideally no – sealed brood cells in the colony. In southern England, the best time to treat is mid-December to early January. Even

ABOVE Trickling oxalic acid.

though it may be cold at this time, you can open hives quickly to check the few frames containing bees to see if there is any sealed brood. If there is very little, you can treat at that time, or use an uncapping fork to remove the cappings to expose any varroa mites inside.

You should use an approved oxalic acid treatment that usually comes as a powder to be mixed in a sugar syrup. You will need a syringe that can be loaded with enough mixture for at least one colony. Remove the roof and crownboard and quickly dribble 5ml only (or as instructed) of the mixture along the length of the gap between frames where you can see bees. It will trickle down onto the bees and treat the mites on them. Then quickly replace the crownboard and roof. Over the following days, check the mite drop on the inspection tray. If it is high, repeat the treatment after seven days.

ABOVE Wasp traps can be effective but avoid using them if you can as wasps too are ecologically important.

An alternative method involves vaporising oxalic acid and injecting it into the colony as a cloud of gas. This method is slightly more effective against varroa but requires expensive equipment and is potentially more injurious to the user, so tends mostly to be used by large-scale beekeepers.

Emergency treatments

Sometimes in mid-summer you might find that a colony has a very high mite count. It is best to use a rapid mite treatment at this time. Formic acid-based treatments can be used for a treatment that takes 7–21 days and can kill mites below wax cappings. These treatments are ideal for use between the spring and summer flows when there are no supers on the hive. One treatment, MAQS, can be used when supers are on the hive.

Our knowledge of varroa and other bee diseases is developing continually, and the availability, usage instructions and legality of commercial treatment products can change. Check BeeBase and reputable suppliers for up-to-date information, speak to your local bee inspector, and see the recommended Further Reading about bee diseases.

Pests

Wasps

Of the 9,000 or so species of wasps found in the UK, just two can become a serious pest of honey bees. The common wasp (*Vespula vulgaris*) and the German wasp (*V. germanica*) are similar in size and appearance. They are predominantly yellow with black stripes, and are familiar to everyone as the invaders of summertime picnics and barbecues. These wasps catch other insects to feed to their larvae. The larvae then secrete a sugary substance that the adult wasps feed on. In late summer, as wasp nests decline, there are fewer larvae to provide sugar for the large numbers of adults, who then search for alternative sources. Honey is ideal, and beehives are a relatively easy target. Hungry wasps can invade honey bee colonies, robbing them of honey and larvae. A weak colony under sustained attack can be killed in a matter of days.

The best way to avoid trouble with wasps is to keep strong colonies that can defend themselves. Do not give wasps the opportunity to begin exploiting colonies – once they have found their way in, they are hard to stop. In late summer, as soon as you see one or two

ABOVE Bees defending a reduced entrance.

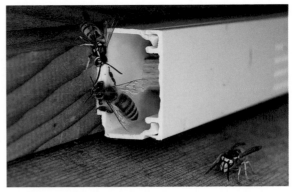

ABOVE Various entrance defences are commercially available.

wasps taking an interest in your hives, reduce entrances dramatically – just one or two bee spaces (a small finger's width) is ideal. Your bees should be able to defend this. Check that all hive parts are securely put together and there are no gaps for wasps to get in.

If wasps continue to be a problem, use traps to divert them. Commercially available traps work well, as do jam jars with 8mm holes drilled in the lids and half-filled with a watered-down mix of beer and jam (never honey). These may need emptying regularly. They are best placed away from hives to draw wasps away from them.

If colonies suffer serious attack, hives may need to be moved away (more than 3 miles/5km) to a safer place. This is more likely to be necessary with nucleus colonies that are small and less able to defend themselves. Nucs under attack can be united with stronger colonies (page 103). Wasps usually only present a serious problem for a few weeks. Their numbers usually decline quite rapidly in autumn and they disappear after the first hard frosts.

European hornets (*V. crabro*) sometimes prey on honey bees but rarely present a serious threat, especially if colonies are strong. These hornets are relatively scarce and have a completely undeserved reputation for being aggressive, so if you see them in your garden, enjoy them for their magnificent beauty. Try to ensure anything being used to trap wasps does not also harm European hornets.

Mice

To mice, a beehive in winter offers a warm, dry, secure home with a built-in larder. In late autumn or early winter, as soon as the nights begin to get cold and damp, protect hive entrances with mouseguards.

The simplest are strips of metal with 10mm holes that bees can get through, but mice cannot. They are held in place with drawing pins. Some hive floors have a permanently mouse-proof narrow entrance so don't need additional precautions. Mouseguards can be taken off in early spring.

ABOVE Mouseguard on a hive entrance in early spring.

ABOVE Winter mouse damage.

Woodpeckers

In winter, green woodpeckers sometimes peck holes in hives and eat bees and larvae. Woodpeckers might ignore hives for years, but once they learn the habit (usually in a severe winter) they can be difficult to stop. If woodpeckers become a problem, hives can be circled with chicken wire for the winter months, the wire touching hives only at the corners. Alternatively, pin heavy-duty plastic sacks to the sides of hives, allowing them to flap at the bottom and preventing woodpeckers getting a grip.

Wax moth

In nature, wax moths perform the important job of cleaning out abandoned or diseased bee nests, creating clean, empty spaces for new swarms to move into. But for bees in hives, and beekeepers, they can be a real nuisance.

There are two species: the greater wax moth (*Galleria mellonella*) and the lesser wax moth (*Achroia grisella*). These cause little trouble in strong colonies, but as soon as comb is left undefended, perhaps in smaller colonies or when stored by the beekeeper, the moths can thrive and make a real mess. The greater wax moth prefers brood comb while the lesser wax moth is more a pest of stored, empty super combs. The larvae chew through comb and create a messy, fluffy nest of silken strands, dotted with faeces. Greater wax moth larvae burrow through occupied brood comb, leaving a trail of partially capped brood cells behind them which can sometimes be mistaken for bald brood (see page 137). When they pupate, greater wax moth larvae eat into hive woodwork and frames to create secure burrows, damaging and weakening equipment. The following measures will help to prevent wax moth becoming a problem:

- Try to keep strong colonies.

- Smaller colonies occupying a few combs should have excess combs removed from the hive.

- Squash moths and larvae if you see them in the hive.

- Avoid leaving pieces of wax or other moth-attracting hive debris on the ground in the apiary.

- Super combs to be stored for winter can be placed in the freezer for 48 hours to kill moth eggs and larvae, before being wrapped in plastic bags or in sealed plastic boxes.

ABOVE Guard hives against woodpeckers – this one has already been pecked at.

ABOVE Brood frames infested with wax moths and their larvae.

BELOW The greater wax moth is the most troublesome of the two species.

- Large quantities of extracted super combs are best stored wet (not returned to the bees for cleaning after extraction) in clear plastic boxes (moths dislike light). Keep them somewhere cold over winter, such as a greenhouse or lean-to shed.

- Combs stored in supers over winter can be treated with acetic acid to kill wax moths. See the National Bee Unit website for instructions.

Exotic pests

Several potentially problematic non-native pests could be a very serious threat to honey bees if they establish themselves in the UK. These species are most likely to be found in or around bee colonies, so it is important that beekeepers know how to recognise them and report suspected sightings as early as possible.

Tropilaelaps mites

Tropilaelaps (*Tropilaelaps* spp.) are two species of mite from Asia where they are parasitic on *Apis dorsata*, the giant honey bee. Like varroa mites, they have jumped species to the western honey bee and could cause serious harm if they reach the UK. As tropical species it is possible they might not establish themselves as successfully as varroa. Tropilaelaps mites reproduce in brood cells like varroa, but they cannot feed from adult bees so are reliant on the presence of brood in a colony to survive.

Tropilaelaps mites are notifiable in the UK, so you must report them if you think you have found them. They are smaller than varroa at around 1mm long and 0.6mm wide, and are a similar reddish-brown colour.

Small hive beetle

Small hive beetle (*Aethina tumida*) originated in South Africa but has found its way to countries including Australia, Canada, Mexico and the USA, where it causes serious damage to honey bee colonies. Eggs laid in hives hatch into rapacious larvae that eat bee eggs, brood, pollen and honey, destroying comb and the whole colony if given a chance. The mature larvae leave the hive to pupate, burying themselves in the surrounding soil before emerging as adult beetles that can fly considerable distances to find and infest other colonies.

Worryingly, small hive beetle has been found in parts of Italy where attempts to control it have been moderately successful. Queen bees bred in Italy are imported to the UK, and this is one way that this serious pest could arrive here. They might also arrive on imported fruit.

The adult beetles are black, 5–7mm long and have distinctive clubbed antennae. The larvae are light brown, 10–11mm long and have three pairs of legs at their head end and spines along their back.

Small hive beetle is notifiable and suspected sightings should be reported immediately.

Asian hornet

The Asian or yellow-legged hornet (*Vespa velutina*) is the most imminent invasive threat to UK honey bee colonies. Native to Asia, it was identified in Europe for the first time in 2004 when found in south-west France, probably having arrived in a consignment of pottery from China. It quickly established and spread to many regions of France, where it has severely affected bees and beekeeping activity. It has since become established in other countries including northern Spain and Portugal as well as the Channel Islands of Jersey and Guernsey.

RIGHT Tropilaelaps mite.

BELOW Small hive beetle and its larva.

BELOW The Asian hornet, a very serious threat.

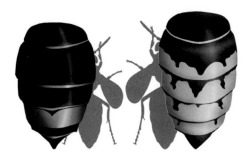

ABOVE Asian hornet (left), European hornet (right). The silhouettes are approximate actual size.

In 2016 the Asian hornet was discovered in the UK for the first time, in Tetbury, Gloucestershire. A huge operation headed by the National Bee Unit searched for and destroyed the nest. Since then, a small number of sightings have been confirmed in most years, the nests being located and destroyed. In 2023, numbers rose alarmingly.

Despite alarmist press reports about 'giant killer Asian hornets' (often accompanied by incorrect photographs), the Asian hornet is smaller than our native hornet. Adult workers are about 25mm long and queens about 30mm. Its abdomen is mostly black or dark brown except for the fourth abdominal segment, towards the rear, which is yellow. Its face is orange with brownish-red eyes.

The most useful distinguishing feature is that it has yellow legs – which is why the name yellow-legged hornet is sometimes used to help aid identification.

Queens hibernate over winter and emerge in spring to seek out sugary food to build up energy. An embryonic nest is built, often low down in a hedge, to accommodate an initially small number of eggs and the subsequent larvae and workers. As the colony grows, a larger nest is built – either around the embryonic nest or elsewhere, often much higher up in a tree.

Colonies increase rapidly, producing up to 6,000 individuals. The adult hornets look for sources of protein to feed their young and, although they may prey on all kinds of insects, beehives are a favoured source. Hovering or 'hawking' near the entrance of a hive, the hornets wait for laden, slower-moving foragers to return, and then swoop down and grab them in mid-air. They decapitate their prey and feed on the flesh of the thorax, feeding some to their larvae.

A single hornet can catch dozens of honey bees each day, and a colony will consume thousands, quickly decimating honey bee colonies.

In autumn, the colony's priorities shift from foraging and nest expansion to producing on average 350 gynes (queens) and male hornets for mating. After mating, newly fertilised queens leave the nest and find somewhere to overwinter. The old queen dies and her colony dwindles to nothing. Successfully overwintered young queens begin the process again the following spring.

For the time being, the best thing beekeepers can do to prevent the Asian hornet from becoming a problem is to keep a careful lookout for them. Beekeepers are well placed to report potential sightings as hornets are likely to target their colonies.

STEPS ALL BEEKEEPERS SHOULD TAKE

- Familiarise yourself with the appearance of the Asian hornet, especially compared with the European hornet.

- Download the free Asian Hornet Watch app to identify and report suspected sightings.

- Set up bait stations to attract Asian hornets for observation. These are designed to avoid the accidental trapping and killing of beneficial insects.

- Use sugar-based lures in spring and autumn and protein-rich meat or fish-based lures in summer, as directed by the NBU.

- Observe hives in summer and look for Asian hornets and their characteristic hawking actions.

- A children's seaside fishing net is useful for catching a sample should you think you see Asian hornets near hives.

- In autumn, observe flowering ivy and rotting fruit, which are attractive to Asian hornets both for sugar and for insect prey.

- Keep up to date. Advice from the National Bee Unit is likely to change as we learn more about Asian hornets and their lifestyle.

Replacing comb

Old brood comb can harbour the pathogens responsible for various diseases and can store residues from treatments, pesticides and other unwelcome pollutants. Over time, brood comb can become dark, damaged and misshapen, leaving combs weaker and with fewer cells available for raising brood.

To keep colonies healthy, it is recommended that brood comb is regularly replaced. The easiest way to do this is to have a programme of ongoing replacement, changing about one-third each year. During inspections, take note of any frames of comb that are particularly old, dark or in poor condition. Throughout the season, gradually move the selected combs towards the edge

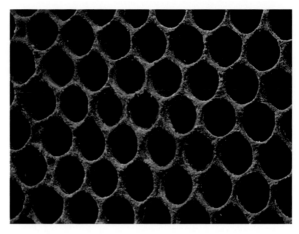

ABOVE Old, dark comb can impact colony health.

of the nest, being careful not to create any gaps in the brood. The selected combs will be ready for replacement with new frames of foundation the following year, once conditions allow the bees to draw new comb.

Shook swarms

If a colony has a brood box full of old, dark brood comb, you might consider replacing all of it at once by the shook-swarm method. This is also a good way to move a colony into a clean hive so that the old one can be cleaned or repaired.

Shook swarms involve shaking all of the bees in a colony into a new, clean hive that is filled with frames of foundation or clean, drawn comb. This is a disruptive process for a colony as it involves sacrificing all of the existing brood and stores but, if done to a healthy colony in late spring or early summer, the bees will make new comb and rear replacement brood amazingly quickly. Shook-swarmed colonies need significant feeding to help them to build new comb and to replace the stores you have removed.

Shook swarms are also a method of controlling moderate cases of European foulbrood (EFB), but should only be used for this purpose in consultation with your bee inspector. Shook swarming will remove all reproducing varroa mites in brood comb – about 70 per cent of the varroa likely to be in a colony. An oxalic acid-based treatment used within a few days of the operation will kill many of the remaining adult varroa mites which will be living on the bees.

ABOVE Brood frames, in their first, second and third years, showing deterioration in comb quality.

How to shake frames of bees

The comb change technique described below requires bees to be shaken from their frames. The technique is also useful any time you want to carefully inspect frames. If done well, this will not harm the bees and they will fall into the hive quickly, reorganising themselves rather than flying up to harm you in any way. To shake bees from frames, they need to be held slightly differently. Instead of holding the lugs of the frame as you might during a normal inspection, wrap your hands around the sides of the frame so that your fingers are underneath the lugs. Then curl your thumbs over the top of the lugs. This way you can give the frame a firm shake and there will be no chance of dropping it. Lower the frame about half-way down into the hive and give it a very strong downward shake. Do this by slowly raising it by about half the frame's height before very rapidly thrusting it downwards, stopping the movement with a sharp jolt. One very firm shake will dislodge most of the bees and they will fall in a heap on the floor of the hive. If lots remain, give the frame another shake. You will hear a roar from the bees as they fall into the hive, and then they will begin to spread across the floor and up onto the frames of foundation, where most of them will stay. Gently brush any remaining bees from the frame into the box.

Shook swarm technique

Shook swarms are best performed on a fine day in late spring or early summer, when the bees are flying well and many of them will be away from the hive.

WHAT YOU WILL NEED

- A complete, clean hive: floor, brood box, queen excluder, crownboard, super and roof
- A queen cage
- A large-capacity rapid feeder
- Enough syrup to fill the feeder (600ml water to 1kg white sugar)
- A lit smoker and water spray bottle
- Bee brush
- A large, sealable container or heavy-duty bags to receive old brood combs

1. Gently smoke the colony and remove the roof and crownboard from the original hive.

2. Go through the brood box and find the queen. You can cage her and put her somewhere cool and dark, or you can put the frame she is on in an empty nuc box with the roof on.

3. Move the original brood box to one side, placing it on top of the removed roof. It might be heavy, so get help with this if necessary. Keep smoking the bees gently now and then if necessary – or a misting of water works well to calm them down.

4. Remove the old hive floor from the stand and replace it with a clean one. Place a queen excluder over the floor. This will make it impossible for the queen to leave the brood box following the completion of the shook swarm. With no brood to keep them in the hive, the colony might otherwise abscond.

5. Add a clean brood box filled with new frames of foundation. Remove three frames from the centre of the box and put them to one side.

6. Now you can perform the shaking part of the shook swarm. The frames need to be firmly shaken to dislodge the bees quickly and with minimum disturbance, as described earlier.

7. Working quickly but carefully, lift a frame of bees out of the old brood box and move it over to the new box. Firmly shake the bees into the gap created in the new hive by removing the three frames.

8. Place the old, emptied frame in a plastic, lidded box, a spare brood box with a lid, or a heavy-duty bag. Seal it to prevent any robbing. As soon as you have finished, it is best to destroy the old frames (see below).

9. Shake all of the bees from all of the old frames and then shake or brush any remaining bees from the sides of the emptied brood box into the new one.

10. Carefully reintroduce the queen, either by releasing her from her cage or by shaking the frame she is on into the new box. If shaking her in, make sure the queen is somewhere in the centre of the frame so that she won't be crushed if you accidentally scrape the sidebars of the frame on the insides of the brood box. Don't be too gentle – it is best to give a firm shake to make sure the queen and her accompanying bees all fall off on the first attempt. Check that you can see her on the floor so that you know she

is definitely in the brood box. She will probably make a dash for the shade of the nearest frames and climb between them.

11. Gently place the three removed frames of foundation back into the gap in the new brood box, and add the clean crownboard.

12. Add an empty super and a feeder. Give as much syrup as the feeder will take and keep it topped up over the next couple of weeks. Reduce the entrance of the hive to minimise the risk of robbing.

13. Check the colony after a week. You should find that plenty of comb has been drawn and that the queen has started to lay again. Once she is laying well and some brood is capped, you can remove the queen excluder from beneath the brood box. Keep supplying with syrup until all of the comb is drawn and there are enough stores to keep the colony going until the next flow.

ABOVE Adding syrup to a large-capacity feeder.

ABOVE New, clean brood comb under construction.

By performing a shook swarm, you will have transferred your colony onto all new frames of foundation and, before long, freshly drawn comb. Removing and destroying many frames of brood can seem an awful waste, but the bees respond to shook swarming with great vigour and very quickly replace the brood, often going on to become strong and productive colonies later in the same season.

Bailey comb change

The Bailey comb change (named after Dr Leslie Bailey, the British bee researcher who developed it) is a gentler way of moving all of your bees onto new comb. It is a good technique to use if you have a weaker colony, perhaps one that is failing to build because of nosema. It's a good system if you are daunted by the idea of vigorously shaking frames of bees, but it does take much longer than the all-at-once efficiency of the shook swarm.

Several variations of the Bailey comb change can be deployed, depending on the vigour of your colony and on your particular situation, but below is the basic principle:

WHAT YOU WILL NEED

- An additional brood box
- Frames of foundation
- Rapid or other feeder plus sugar syrup
- An empty super and a second crownboard
- A Bailey eke (see below)

Remove all of the peripheral frames of stores from the brood box, leaving only the occupied brood frames with dummy boards enclosing them.

1. Shake the bees from the removed frames onto the remaining combs – making sure the queen is safely in the brood box.

2. Place a second, clean brood box over the original. It should contain frames of foundation which correspond in number and position to the frames below them. Again, use dummy boards to sandwich these frames. The bees will move up from the old frames onto new ones directly above.

3. Place a crownboard over the top brood box, followed by an empty super. Add a feeder and sugar syrup. Close the hive.

4. Once the foundation in the top chamber has been drawn into new comb (about a week), move the queen up onto the new comb and place a queen excluder between the two brood boxes.

5. Close off the original hive entrance and add a new entrance, above the queen excluder but below the top brood box, by inserting a Bailey eke. This is an easily made frame of wood about an inch high with an entrance hole. Flying bees will learn to use the new entrance.

6. As the brood in the lower box emerges, all of the workers will migrate to the new comb in the box above, where the queen will have started laying. Once the old comb in the lower box is free of brood, the box and its old comb can be removed. Then place the upper box on a new, clean floor in the same apiary position and add additional frames of foundation. You will now have a clean hive with fresh comb. The whole process should take about a month.

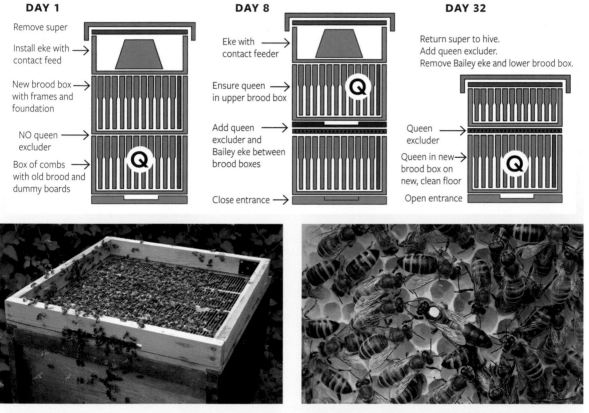

DAY 1

Remove super

Install eke with contact feed →

New brood box → with frames and foundation

NO queen → excluder

Box of combs → with old brood and dummy boards

DAY 8

Eke with → contact feeder

Ensure queen → in upper brood box

Add queen → excluder and Bailey eke between brood boxes

Close entrance →

DAY 32

Return super to hive.
Add queen excluder.
Remove Bailey eke and lower brood box.

Queen → excluder

Queen in new→ brood box on new, clean floor

Open entrance

ABOVE A Bailey eke and entrance on top of a brood box.

ABOVE The queen should soon be laying in new, clean comb.

A clean start

Shook swarms and Bailey comb changes are best done relatively early in the season so that colonies have the best part of the spring and summer to expand. By the end of the season, you will have a colony housed on good-quality, clean frames and comb. You will not need to replace any comb the following season, but can begin an ongoing programme of quality control – aiming to replace about a third of the darkest or most unsatisfactory combs each season thereafter. There is great pleasure in knowing that your bees are using new, brightly coloured wax comb. It is more pleasant for you to work with, and makes it much easier to observe eggs and larvae in the cells.

The beekeeping year

Season
MID-WINTER
(January)

Approx size of colony
10,000

Colony activity
- Queen might lay eggs in small numbers
- Small quantities of pollen brought in
- Colony clustered much of the time
- Throwing out dead bees

Jobs for beekeeper
- O. acid varroa treatment if not already done
- Heft to check stores
- Clear entrances of dead bees
- Clear snow from entrances
- Be vigilant for woodpecker damage

Significant forage
- Hazel
- Snowdrops

> Months given are approximate, exact times can vary according to climate and location.

Season
LATE WINTER
(February)

Approx size of colony
8,000

Colony activity
- Collecting water to dilute stores
- Increased pollen foraging on warm days
- Queen's laying increases

Jobs for beekeeper
- Too early to open hives. Read last year's records
- Heft to check stores, feed fondant if necessary
- Survey local fields for growing oilseed rape
- Watch bees at entrances – pollen coming in is a good sign
- Check stocks of frames, foundation, etc. Make spring shopping list

Significant forage
- Alder
- Hazel
- Snowdrops
- Winter aconite

Season
EARLY SPRING
(March)

Approx size of colony
5,000

Colony activity
- Regularly flying
- Bringing in pollen and nectar
- Brood nest expanding
- Might throw out old pollen

Jobs for beekeeper
- Keep checking stores
- Make frames and prepare supers ready for flow
- Monitor varroa drop
- Visit spring beekeeping events – learning and shopping

Significant forage
- Crocus
- Hellebores
- Wild cherry
- Willow

Season **MID-SPRING** (*April*)	Season **LATE SPRING** (*May*)	Season **EARLY SUMMER** (*June*)
Approx size of colony 12,000	**Approx size of colony** 30,000	**Approx size of colony** 55,000

Mid-spring (April)

Approx size of colony

12,000

Colony activity

- Queen should be laying very well
- Brood nest expanding, stores depleting
- First drones usually raised
- Queen cells now could mean queen supersedure

Jobs for beekeeper

- Regular inspections begin
- Full disease check on all brood frames
- Safe to feed syrup if stores are low
- Begin looking for swarm preparations
- Revise swarm management plans
- Possibly add first super

Significant forage

- Blackthorn
- Crab apple
- Dandelion
- Field maple, sycamore
- Oilseed rape
- Orchard fruit (plum, pear, cherry)

Late spring (May)

Approx size of colony

30,000

Colony activity

- Queen laying very heavily
- Colonies likely to think about swarming
- Honey being made from spring flowers
- Young bees want to make wax – give them frames if possible

Jobs for beekeeper

- More supers added
- Check brood comb condition – do any need replacing?
- Be on high alert for swarming; equipment ready
- Split large colonies for increase or swarm prevention

Significant forage

- Hawthorn
- Apple
- Field bean
- Horse chestnut
- Oilseed rape
- Dandelion

Early summer (June)

Approx size of colony

55,000

Colony activity

- Colony population nearing peak
- Possible swarm preparations
- Foraging whenever possible
- Drones flying, new queens mating

Jobs for beekeeper

- Extract spring honey crop
- Assess quality of queens – do they need replacing?
- Split colonies to make new queens
- Treat for varroa mites if needed between spring and summer crops
- Look for signs of starving during June gap

Significant forage

- Bramble
- Field bean
- White clover
- Phacelia
- Lime

Season
MID-SUMMER
(*July*)

Approx size of colony
60,000

Colony activity
- Main summer honey flow
- Queens sometimes go off lay briefly
- Amount of brood begins to decline
- Swarming becomes less likely

Jobs for beekeeper
- Give plenty of room in supers
- Ease up on inspections unless swarming is thought likely
- Prepare for extraction of honey
- Source new queens if you haven't raised your own

Significant forage
- Bramble
- Borage
- Garden flowers
- Lime
- Sweet chestnut
- White clover

Season
LATE SUMMER
(*August*)

Approx size of colony
40,000

Colony activity
- Queen might have second laying peak, producing bees for winter
- More propolis used around hive
- Possibility of robbing if nectar flow is poor

Jobs for beekeeper
- Unite colonies using new queens
- Treat for varroa mites immediately after honey harvest
- Reduce hive entrances and watch for wasps
- Inspect all brood frames for disease

Significant forage
- Bramble and thistle
- Goldenrod
- Himalayan balsam
- Ling heather in some areas
- Rosebay willowherb
- White clover

Season
EARLY AUTUMN
(*September*)

Approx size of colony
25,000

Colony activity
- Frantic foraging if weather is good and ivy flowering
- Brood nest starts to shrink; combs backfilled with honey
- Fat winter bees emerging
- Remaining drones may be expelled from nest

Jobs for beekeeper
- Monitor stores in brood box
- Remove remaining varroa treatments; make legal records
- Feed thick syrup or fondant if necessary
- Last chance to unite colonies before winter
- Watch for hawking Asian hornets

Significant forage
- Himalayan balsam
- Ivy

Season	**Season**	**Season**
MID-AUTUMN	LATE AUTUMN	EARLY WINTER
(*October*)	(*November*)	(*December*)

Approx size of colony

15,000

Colony activity

- Queen's laying is significantly reduced
- Colony activity is significantly reduced
- Bees begin to cluster on cold nights
- Summer bees dying off

Jobs for beekeeper

- Feed fondant if you are worried about low stores
- Make sure hives and stands are secure for winter
- Fit mouseguards to hives
- Strap hives down for winter

Significant forage

- Himalayan balsam
- Ivy

Approx size of colony

12,000

Colony activity

- Rare flights on warmer days
- Queen probably stops laying
- Colony clustered to keep warm

Jobs for beekeeper

- Leave bees alone
- Clean, fix and store equipment
- Tidy apiary, cut back grass or branches

Significant forage

None

Approx size of colony

10,000

Colony activity

- Very little

Jobs for beekeeper

- Treat for varroa mites using oxalic acid
- Give friends and relatives a gift of home-produced honey
- Process beeswax. Make candles, wax, wraps and lip balms

Significant forage

None

PART TWO

Bees in your garden

Of the 20,000 or so known species of bees in the world, just a handful are classified as honey bees. The vast majority are solitary bees, which neither live in colonies nor make any honey. Around 250 species are bumblebees, whose lifestyle sits somewhere between that of honey bees and solitary bees. What sets most bumblebees and solitary bees apart from honey bees is that they cannot be kept in hives, bred, fed, or treated for disease. These are our wild bees, and although we have almost no close contact with them, their survival now rests on our ability to care for them and the environments that they depend on.

Bumblebees

Bumblebees are the charismatic larger cousins of the honey bee and are generally more noticeable as they busily navigate our summer hedgerows, fields and gardens. They attract much admiration and affection for their fluffy, often colourful appearance and their melodious humming, which for many is the quintessential sound of summer. The Old English name for bumblebees, and a few other large, flying insects, was 'dumbledore' – something perhaps of interest to fans of Harry Potter. Charles Darwin and many early entomologists knew them as 'humblebees', a reference to the noise they make. The name bumblebee had been around for some time, but seems only to have gained popularity and then permanence after another Potter – Beatrix this time – created a troublesome bee character called Babbitty Bumble for *The Tale of Mrs Tittlemouse* (1910).

Both honey bees and bumblebees are in the subfamily Apinae, but honey bees are in the genus *Apis*, whereas the genus of bumblebees is *Bombus* – a reference to the deep, booming sound they make when in flight. There are 24 species of bumblebee in the UK. Some are very common – seven being seen frequently in gardens in most parts of the UK – while others are rare, and some are highly endangered. Two UK species have been recorded as extinct in the last 80 years and efforts are being made to conserve a further eight species – though the outlook for some is very uncertain.

In some respects, bumblebees are quite like honey bees. Most species are social, living in colonies headed by a single mother queen. Worker females collect pollen and nectar which is brought back to feed young in the nest. Females can also sting – though they are much less likely to do so than honey bees. There are significant

LEFT Male common carder bumblebee (*Bombus pascuorum*).
PREVIOUS PAGES UK bee species vary enormously. Here a buff-tailed bumblebee shares a nasturtium with a common yellow-faced bee.

differences too, however. Bumblebee colonies are not perennial; they do not make honey to help them survive the winter because the entire colony will die at the end of summer – the only survivors being new, young queens which hibernate alone. And not all bumblebees live in cooperative harmony; some are cuckoos, hijacking the nests of other species, usurping the rightful queen and forcing the workers to raise their own young instead.

The bumblebee life cycle

Like honey bees, bumblebees have three castes – queens, workers and males – although there are some exceptions as explained below. Bumblebee queens are the only members of a colony to live through winter. They hibernate somewhere likely to stay dry and cold. This might be behind ivy on a north-facing wall, in a pile of logs, under a shed, or below ground in an abandoned mouse burrow. Queens emerge again in spring, with some species becoming active earlier than others. In the southern UK they are typically seen from about late February. With the warming climate, emerged queen bumblebees are now frequently seen several weeks earlier than was once the case; some possibly even nest and rear young in the winter months – but they need suitable winter-flowering plants if they are to survive.

Once active, overwintered queens must first find a source of nectar to give them energy to keep warm and stay mobile. They also eat some pollen to help the growth and maturation of their ovaries. Their next priority is to establish a nest. Different species show a preference for different kinds of nesting site. Some prefer to be underground (perhaps in an old rodent's nest), others at ground level (in thick grass or compost heaps) and others high up (in a tree cavity or bird box). Some species, such as the garden bumblebee (*Bombus hortorum*), can be quite eccentric in their choice of nest site, and have been found making their home in

ABOVE Hosepipe entrance to an artificial nesting chamber being used by red-tailed bumblebees.

lawnmowers, watering cans, and even in the pocket of a coat hanging in a shed. Gardeners will be familiar with the sight and sound of large bumblebee queens zigzagging their way across the garden on sunny, early spring days, stopping frequently to inspect possible nesting sites and disappearing underground to review

what they have found – they are choosy house hunters. In addition to a suitable space, an existing supply of nesting material is required, which could be dried grass or leaves, fine roots or moss – often remnants of a rodent's nest.

With a nest established, queens, rather charmingly, sculpt a small thimble-like wax honey pot. Bumblebees don't actually make honey, but will fill their pot with nectar, a guaranteed source of energy should the unpredictable early spring weather keep them at home for an extended period. Next, they gather pollen in much the same way that honey bees do, collecting it on their hairs, moistening it with nectar and bringing it back to the nest on their hind legs to form a ball, which is a larder to feed their young. On top of this pollen store they sculpt wax walls to make a nest, within which the first eggs are laid. A wax roof is then built over the top. Several such chambers will be constructed, the queen lying over them to keep them warm and incubate the eggs within.

The eggs hatch after about four days, the resulting larvae feeding on the store of food on which they lie, as well as a mixture of pollen and nectar regurgitated

ABOVE A buff-tailed bumblebee queen searching for a place to nest.

by the queen. They grow rapidly, shedding their skins several times until after about a week they pupate, spinning a silk cocoon around themselves, inside which they will go through metamorphosis.

Between three and four weeks after eggs were laid, adult bumblebees emerge. These will be sterile worker females whose job it will be to gather and bring back to the nest further supplies of pollen and nectar, to support the raising of more young. At first, the queen might continue foraging too, but when her workforce of daughters reaches a sufficient size she will retire, staying in the nest to concentrate on laying more eggs.

The first workers to be born each year are often very small, a result of the limited food and attention given to them as larvae. In late spring these mini-sized bumblebees can be very apparent as they forage frantically on flowers, a heart-warming sign that nearby bumblebee nests are thriving. As more workers become available to serve the nest, the quantity and quality of available stores increases and subsequent generations grow larger until, by mid- to late summer, worker bees can be quite substantial. The size of the colony grows, too, depending on the species. Buff-tailed bumblebee (*Bombus terrestris*) colonies can grow to include up to 400 individuals, whereas early bumblebee (*B. pratorum*) colonies might have only 50.

At some point in late summer, the queen lays eggs that will produce males and new queens. When mature, these do not collect resources for the colony but forage for pollen and nectar to build their own strength – although some young queens will bring resources back to the colony until they are ready to leave the nest permanently. When sexually mature they will mate, most queens mating just once with a single partner. Unlike the aerial antics of honey bees, bumblebees like to keep their feet on the ground; coupling pairs can sometimes be found on lawns or perched on walls or branches. Mated queens then search for a place to hibernate, and soon settle down in the place where they will spend the winter. The males do not return to the nest, but spend the rest of the summer flying around rather listlessly and feeding somewhat half-heartedly, often sleeping inside flowers, until eventually they die. The old nest gradually peters out too, with workers and the old queen dying off once their job of producing new queens and securing the future of their species has been achieved.

Cuckoos in the nest

Six of the 24 species of UK bumblebee are cuckoos, also known as brood parasites or inquilines. These species emerge from hibernation later than other bumblebees, having waited until their target species have established their nests. Each species of cuckoo bumblebee has at least one specific host, and has similar markings to the host to assist it in its skulduggery. There are various physical differences, however. For example, cuckoo bees never collect pollen for transportation to a nest, so they do not have pollen baskets on their back legs. They have darker-coloured wings with more prominent veins and are less hairy, revealing the body armour used to protect them when forcing their way into a nest. When foraging on flowers they can be slow-moving and show a reluctance to fly, unlike other bumbles who move energetically from flower to flower. When they do fly, their buzz is softer and lower-pitched, and this

BELOW A male Southern cuckoo bee (*Bombus vestalis*) foraging on *Knautia macedonica*.

is the reason why they were originally assigned their own slightly sinister-sounding genus name, *Psithyrus*, meaning whispering. They are now recognised as true bumblebees and have been reassigned to the genus *Bombus*.

It is thought that the female cuckoo bumblebee uses various mechanisms to successfully invade nests, including entering in stages in order to gradually acquire the correct colony odour. Once accepted within the nest, she kills or subdues the genuine queen, destroying her eggs and replacing them with her own before assisting or cajoling the workers of the nest to raise her young. Cuckoos have only two castes – queens and males. There is no need for any workers since all the work is done by others. Young cuckoos can continue to be produced until the host nest begins to falter – without its own queen, no new young workers will be produced after the cuckoo has invaded and so nests soon run out of steam. Mature cuckoo queens and males leave the nest and mate much as other bumblebee species do, the males later dying and the queens going into hibernation over winter.

The idea of cuckoo bumblebees horrifies some people and can be at odds with their idea that all bumblebees are cuddly and benevolently industrious. However, there is no evidence that they adversely affect populations of their host species. As long as the overall bumblebee population is healthy, the proportion of nests invaded by cuckoos seems to remain relatively stable and both hosts and cuckoos are able to survive in a balanced way.

Identifying bumblebees

It's rarely difficult to find a bumblebee in summer; you often hear them before you see them. Once spotted they are relatively easy to keep track of as they move from flower to flower at a brisk but trackable pace. You can even get very close to them without fear of disturbing them or receiving a sting. But if you want to know exactly what kind of bumblebee you are looking at, things get a bit more complicated.

Conveniently, male bumblebees usually have mustaches.
BELOW Male early bumblebee (*Bombus pratorum*).
OPPOSITE Male red-tailed bumblebee (*B. lapidarius*).

ABOVE The brown-banded carder bee (*Bombus humilis*) is one of our rarer bumblebees.

Of the 24 species of bumblebee in the UK, only a handful are likely to be seen in the average flower-filled garden. Other, rarer, species tend to live in isolated areas or only where their preferred conditions or food plants are available in abundance.

Several of the commonly seen species are quite easy to identify accurately, but a few can be frustratingly hard to tell apart – even for experts. Some look remarkably similar; worker buff-tailed bumblebees look almost exactly like white-tailed bumblebees (*B. lucorum*), for example. To add to the confusion, queens, workers and males of the same species can look quite different to one another but very similar to other species. Some species come in several colour and pattern variations, and their fur can bleach in the sun or wear thin, bestowing quite a different appearance with age. There is also variability in size, sometimes even between bees of the same species and caste. The best time to begin identifying species is in spring when only the queens, who are generally easier to identify, are on the wing. As the season progresses, you can learn to identify the workers and

males as they begin to appear. If you really want to know what species you're looking at, it can be worth taking a photograph and then referring to one of the comprehensive guide books (see Further Reading, page 297) or submitting your picture to an online forum. You can also catch bees in a sweep net and transfer them to an examination container for a closer look. Unless experienced, disturbing them in this way is best avoided unless you absolutely need to know what you are looking at – and, even then, a definitive answer is sometimes only possible if you are prepared to kill and dissect the object of your interest.

There follows a description of the seven most common UK species, along with their associated cuckoos where appropriate. As well as describing the colour, size and habits, tongue length is mentioned: the length of a bumblebee's tongue affects what flowers it is likely to be seen visiting. Along with each description are three diagrams showing the common colour patterns of queens, workers and males. These should help with identification, but remember that there can be some considerable variability.

Buff-tailed bumblebee (*Bombus terrestris*) [Short-tongued]

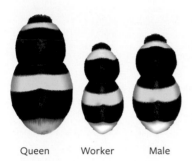

Queen   Worker   Male

Queen

Worker

Among the most common and widespread of UK bumblebees, the buff-tailed bumblebee has the classic black-and-yellow striped appearance that most people associate with bumblebees – along with a tail tip that looks as if it's been dipped in paint. Queens are easy to identify, especially in early spring when they are usually the most frequently observed species, often seen and heard zigzagging low over the ground looking for a nesting site. They are very large and glossy black, with two golden or dusky yellow bands and an off-white, buffish tail. Workers and males look similar but smaller and their tail is pure white, which unhelpfully makes them look almost exactly like worker white-tailed bumblebees; the two species can be almost impossible to tell apart even for experienced bee spotters. Buff-tailed workers often have a very narrow brownish band between the black hair of the body and the white of their tail – a useful distinguishing feature.

Buff-tailed bumblebees nest underground, beneath patios, among tree roots and often in deep rodent or rabbit burrows – nipping in and out of what should clearly be someone else's front door. Their colonies are large, holding up to about 400 bees at the height of summer. Nests can be taken over by the southern cuckoo bumblebee (*B. vestalis*), which is a commonly seen species and often misidentified as the buff-tailed bumblebee.

Buff-tailed bumblebees are short-tongued, so mostly visit flowers with shallow, easily reached nectaries. However, they are expert nectar robbers and will bite a hole at the base of tubular flowers and poke their tongues in to reach the nectar directly. Other short-tongued bees such as honey bees also use the holes, meaning the flower might surrender its nectar without ever being pollinated.

White-tailed bumblebee (*Bombus lucorum* agg.) [Short-tongued]

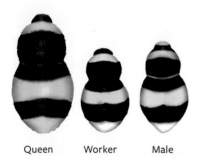

Queen   Worker   Male

Queen

Worker

Male

This is one of the most widespread bumblebees, reaching to the far north of the UK where even the more common and closely related buff-tailed bumblebee is not found – though it is usually not quite as frequently sighted. While workers and males of both species are remarkably similar, the large queens are more easily distinguished, by the pure white tail for which they are named. Their two yellow bands are a paler lemon yellow than those of buff-tailed queens. Males and workers also have white tails but can be hard to distinguish from those of the buff-tailed bumblebee, although males have a hairy yellow moustache on their face and, though generally dark all over, can sometimes be quite yellow in appearance.

The white-tailed bumblebee is an aggregate species, meaning it is a group comprising three very similar but genetically different species. As well as *Bombus lucorum*, the common name also covers *B. cryptarum* and *B. magnus*. The differences are so subtle that they can only be uncovered using DNA analysis, so they are known as cryptic (hidden) species. To anyone other than bee scientists, the name white-tailed bumblebee refers sufficiently to bees of this type and appearance.

Like their buff-tailed cousins, white-tailed bumblebees have short tongues and are very keen on biting holes in tubular flowers in order to rob nectar. They nest in similar underground locations, showing a distinct preference for setting up home under wooden sheds. Nests can be colonised by the gypsy cuckoo bumblebee (*B. bohemicus*).

Garden bumblebee (*Bombus hortorum*) [Long-tongued]

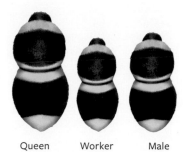

Queen   Worker   Male

This is another classic yellow-and-black striped species that superficially looks very similar to both the white- and the buff-tailed bumblebee. Queens, workers and males all have a white tail but have three yellow bands rather than two. Males sometimes also have some yellow on their head.

The main distinguishing feature of this species is its elongated horse-like head, which is most noticeable as it faces forward during flight. Garden bumblebees have the longest tongue of almost any UK bumblebee (up to 15mm) and sometimes fly with it hanging below them like an elephant's trunk. The long tongue allows nectar to be accessed from the front of deep, tubular flowers like honeysuckle, toadflax, comfrey, penstemons and broad beans. The best way to practise identifying this species is to watch a stand of foxgloves or white deadnettle; the bees are bound to be seen coming and going with their tongues hanging out.

Garden bumblebees are very widespread but are seen a little less frequently than some of the other common bumblebees. Queens emerge from hibernation later in spring than most species, and their relatively small-sized colonies typically start to die out a little earlier than those of some others.

Nests are usually underground with a short tunnel but are sometimes just above ground level, even having been found in old sparrows' nests and tit boxes. Nests are parasitised by Burbut's cuckoo bumblebee (*B. barbutellus*).

Worker

Early bumblebee (*Bombus pratorum*) [Short-tongued]

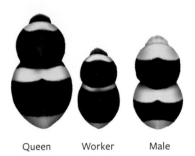

Queen   Worker   Male

Worker

This widespread and common bumblebee is one of our smallest species but more than makes up for its size with its delightful appearance. Early bumblebee queens have two wide, yellow bands and a tangerine-coloured tail that bestows a slightly exotic appearance. Workers and males look similar to queens, although the male's yellow bands are even wider and he has a smart yellow moustache on his face, giving him a very yellow appearance overall. The worker's second yellow band, on the thorax, can be very thin or almost absent.

Sometimes called the early-nesting bumblebee, this is one of the first to establish its nests each year and to finish its breeding cycle. The small-sized colonies are often on the decline by mid-summer and sometimes the bees are not seen beyond late summer, when other species are still very active.

Early bumblebees have short tongues so are often seen on composite flowers (flowers with numerous shallow florets) like daisies, dahlias and knapweed. Early-emerging queens are often seen on willow, rosemary and flowering currant.

This bee tends to favour urban or suburban sites with access to a wide variety of garden plants, and is less common in areas of intensive agriculture. Nest sites are varied and can be just below or above ground, in tit boxes, old bird nests or even as high as roof spaces in buildings. Nests can be invaded by the forest cuckoo bumblebee (*B. sylvestris*).

Male

Common carder bumblebee (*Bombus pascuorum*) [Medium-long tongued]

Queen   Worker   Male

Queen

Very frequently seen and easy to identify, the common carder bumblebee is a gingery-brown colour with varying patterns of light and dark bands on the abdomen. There are quite a few colour variations, ranging from sandy-brown to vivid orange. Although there are two other similarly coloured carder species – the brown-banded carder (*B. humilis*) and the moss carder (*B. muscorum*) – if you see a bumblebee with these colours and markings in your garden, it is almost certain to be the common carder bumblebee.

This petite bee is always a delight to watch as it moves busily from flower to flower, its coloration and furry appearance making it reminiscent of a flying teddy bear. It has a medium- to long-length tongue and favours flowers that are slightly too deep for the shorter-tongued species, such as clovers, vetches, catmint, lavender and green alkanet. Queens appear slightly later than those of most other species, but nests are long-lived and young bees can still be seen on the wing well into autumn, along with straw-coloured, sun-bleached veterans.

Carder bees are named for their habit of carding or weaving grasses and mosses together to make their nests, which are at or slightly

Worker

below ground level, usually among tussocky grass, matted plant roots, or piles of twigs and leaves. The field cuckoo bumblebee (*B. campestris*) attacks the nests.

The other carder species, which include the shrill carder (*B. sylvarum*) and the red-shanked carder (*B. ruderarius*), are among the rarest bumblebee species in the UK.

175

Red-tailed bumblebee (*Bombus lapidarius*) [Short-tongued]

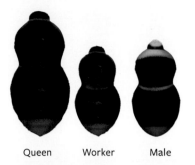

Queen   Worker   Male

Queen

Worker

Male

Its glossy, black velvety fur and a bright, orange-red tail make this one of the easiest bumblebees to identify. It is very common in much of the UK but less so from Cumbria northwards, particularly in more exposed areas – though climate change seems to be aiding its spread northwards.

The queens are among the last to emerge in the spring, but they are large and very noticeable when they do make an appearance. Workers look identical to queens but are smaller. Males, which can be very abundant in late summer, have one or two yellow bands of varying width and shade, and yellow facial hair, making them very attractive but sometimes easy to confuse with male early bumblebees.

Red-tailed bumblebees have short tongues so visit open, easy-to-access flowers, particularly legumes. They love white clover and, perhaps with the exception of honey bees, will be the most commonly seen bees on a clover-rich lawn, moving balletically around the flowers that bend and bob under their weight. They also show a strong liking for yellow flowers like bird's-foot trefoil, horseshoe vetch and oilseed rape, and for flowering trees such as laburnum. This is a warmth-loving species and it can disappear for days at a time if temperatures cool, seeming to appear in abundance again when the sun returns.

This species nests both above and below ground wherever it can find suitable spaces, perhaps behind stones in a garden rockery or inside drystone walls. Its cuckoo species is the red-tailed cuckoo bumblebee (*B. rupestris*), which looks very similar so can easily be confused with its victim.

Tree bumblebee (*Bombus hypnorum*) [Short-tongued]

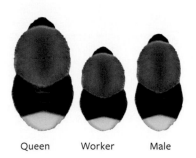

Queen   Worker   Male

Queen

This is a newcomer to the UK from mainland Europe, having first been recorded in 2001 in Hampshire. Since then, it has spread rapidly and can be found everywhere except the far north, although it will likely soon reach there too.

Tree bumblebees are easy to identify. Queens and workers have a thorax covered in rich mahogany-orange coloured hair, and a black abdomen with a white tip to the tail. Males are similar, but the brown colouring can extend to parts of the abdomen.

Tree bumblebees are so named because they nest high off the ground, often in tree cavities or among insulation material if they can find their way into a loft. Another favourite location is bird nesting boxes, even forcing out the established feathered homeowners. They often nest in close proximity to one another, with up to 250 nests per square kilometre having been recorded. When new queens are about to emerge from the nest, males from other nearby colonies buzz excitedly around the entrance, ready to make their advances. These clouds of bees are often reported to beekeepers as swarms of honey bees, and are the cause of many wasted

Nest

journeys. This is the only bumblebee species that displays any kind of seemingly unprovoked defensive behaviour, occasionally bothering and even stinging people who get too close to their nest. But this is rare and a small price to pay for the presence of this otherwise harmless and welcome immigrant bumblebee.

The tree bumblebee is an early-emerging short-tongued species that visits a very wide range of open, shallow flowers. It is thought that the forest cuckoo bumblebee (*B. sylvestris*) might attack the nests of this species. This does not seem to happen frequently, which might account in part for the rapid spread of the species.

Nesting sites for bumblebees

Most gardens can easily accommodate a bumblebee colony or two, but introducing homemade or commercially available nest boxes is a hit-and-miss affair, with even the better designs being ignored more often than not. If you want bumblebees to nest in your garden, your efforts are perhaps best spent creating a range of informal possible nest sites in the hope that some will find favour.

Knowing what attracts bumblebee queens is helpful when designing potentially attractive nest sites. Most species require a ready-made dark, dry cavity. Some favour ground or above ground-level cavities (such as *B. pratorum* and *B. hortorum*), while others prefer to nest below ground (*B. terrestris* and *B. lucorum*). The volume of the cavity can range from the size of a tennis ball to a shoe box. Old rodent nests or tunnels are probably the most frequently used naturally occurring locations, and in wilder areas unused rabbit, badger and fox holes are popular. Holes in and around the roots of long-established hedgerows and large old trees often find favour. Fallen trees with a maze of exposed roots and tunnels can become the equivalent of a bumblebee housing estate.

Queen bumblebees generally like to land on the ground and then walk down into their nest via a tunnel. Tunnels can be quite short or several metres in length. Queens prefer a ready furnished property with various nesting materials that they can rearrange to their satisfaction. If you hope to make a potential nest site

BELOW Queen buff-tailed bumblebee investigating possible nesting sites among moss.

alluring, soft, dried leaves and moss, downy feathers and untreated kapok (sold as stuffing and pet bedding) can all be used. Don't use cotton wool, which is too fine and seems to get caught in the bees' feet. A small amount of dried grass can be added, but it should be very thin and soft, not coarse and spiky like most hay. Try to arrange the nesting materials so that there is a clear passage through them from the entrance tunnel, leading to a snug internal void where a nest can be started.

Pre-existing cavities

Bumblebees are opportunistic nesters, mostly using ready-made cavities that don't require any further modification. Indeed, unlike many solitary bees, bumblebees are poor diggers and are unable to burrow holes and tunnels of their own. They will often use pre-existing cavities which are part of the artificial structures that they come across. This can include squeezing through airbricks to nest under houses, between paving slabs to nest beneath patios, through crumbled pointing to get inside walls and, frequently, beneath the wooden floors of garden buildings. One year I had three white-tailed bumblebee nests under the floor of my garden office, making the floor vibrate gently as the inhabitants buzzed.

If you are willing to offer any such structures to the bees, ensure that access is left unblocked for queens to find their way in. If you have a patio, you could lift up one of the edging slabs and excavate a tunnel and cavity beneath it. In the case of sheds sitting on a wooden base, you could drill one or two entrance holes (about 15mm in diameter) through the timber at ground level and poke some nesting material through the hole.

Creating nesting opportunities

There are lots of ways to create potential bumblebee nesting sites around the garden. Pile fallen leaves and small twigs at the base of hedges to create an understorey that might be used for nesting – the cover may also attract voles, whose burrows and runs make good bee nesting cavities. Piles of unused materials can present nesting opportunities; bricks, roof tiles, terracotta plant pots, logs, branches and pieces of timber will all do. Pile these up in disused, dry and sheltered corners of the garden and try to arrange them so that there are various internal cavities with ground-level entrance tunnels leading to them. Place them against walls, fences, sheds or hedges, as queens tend to fly along linear structures when looking for a nest site. Before moving the materials into position, dig a good-sized hollow in the ground. Put some sharp gravel or small twigs at the bottom to provide drainage, followed by a floor of dry soil and then a handful of nesting material. When the cavity has been covered, make an entrance tunnel by poking a stick through the soil into the void, or lay in a length of ordinary 19mm/¾ inch garden hosepipe. Shelter the tunnel entrances so that rain doesn't drain into the nest. Arranging the hosepipe so that there is a slight rise in the middle of the length will prevent water draining into the nest, or the tunnel flooding mid-way and trapping bees inside.

Rockeries and bee banks (see page 206) can present good nesting opportunities. Dig nesting cavities under large rocks or logs and use a length of garden hosepipe as an entrance tunnel that emerges on a nearby bank. Bumblebee queens particularly explore slopes for potential nesting sites, and can often be seen searching hedge banks in spring. Piling moss around hosepipe entrances seems to make them more alluring to queens. I have seen red-tailed bumblebees (*B. lapidarius*) using rockery nests like this.

Compost heaps are often used, perhaps because they are usually warm, but also because rodents nest and burrow in them, making cavities suitable for bumblebee nests. I once found a white-tailed bumblebee nest inside a butternut squash in a compost heap – the inside had rotted away, leaving behind the dried, shell-like skin of the squash, which made a perfect cavity for a nest. For bumblebees to get access to them, compost heaps need to be open-fronted or have slatted sides, such as those made from pallets. Look out for bumblebee activity in spring and summer so that you don't disturb nests when digging out compost.

High-rise and ground-floor nesters

The tree bumblebee (*B. hypnorum*) routinely nests at height, often in tit boxes but also in tree cavities and lofts. If you want to attract this species to your garden, put up tit boxes, and whether bees or birds occupy them they will be a success. Even if birds get there first, tree bumblebees have been known to drive them out.

Carder bees, in particular the common carder (*B. pascuorum*), will nest in gardens if there is an

ABOVE: An above-ground nest of common carder bees, made from grass and moss.

BELOW Tree-bumblebee nest in a bird box.

undisturbed area of long, unmanaged grass. If uncut for several years, matted dry grass and moss will build up at the base of the newer shoots and this may be used for nesting. Leave the corners of lawns unmown, or planted with various wildflowers (see page 243), or leave a metre or two of lawn uncut around the edge of trees or garden buildings. Carder bees make their nests at ground level, or just below, by weaving together moss and grass, so be careful not to disturb or tread on them.

To help bumblebees thrive and possibly multiply in your garden, the best thing you can do is to adopt the broad principles now widely accepted as best for wildlife gardening in general; don't be too tidy, leave some areas relatively undisturbed or completely wild, avoid the use of any garden chemicals and grow a wide range of plants that are beneficial to wildlife. These subjects are explored in Part Three.

Solitary bees

The vast majority of the world's 20,000 known bees are solitary species. As their name suggests, these bees differ from honey bees and bumblebees in that they do not live in social colonies or share duties like building nests, collecting food or raising young. Nor in most cases do they have queens – just males and females. Most solitary bees interact with others of their own kind only when they mate – usually a rather brief one-off encounter. Females construct nests, lay eggs and collect food for their young but their maternal work ends there and the young must fend for themselves, never to meet their parent. Some species nest in aggregations, dozens or hundreds of nests being built in close proximity and appearing to be part of a family group or colony. In fact, these bees nest near to one another probably only because the local conditions are ideal for tunnelling. They may also be attracted by the pheromones emitted by others of the same species. A few species have developed a small degree of social cooperation, usually extending only as far as sharing a nest entrance hole or provisioning one another's egg chambers in an unorganised way with food supplies. Such behaviour is sporadic even among populations of the same species, so they are still classed as solitary bees.

Solitary life cycles

The lifestyles and nesting habits of solitary bees are tremendously varied, and there is not the space here to detail the differing behaviours of the many species. Fortunately, most share some common aspects so the following life cycle description, though rather generalised, will help you to understand what the bees in your garden are doing, and what you can do to protect them and help them to thrive.

LEFT A male hairy-footed flower bee (*Anthophora plumipes*), likely to be one of the first solitary bees seen in the spring, and among the easiest to identify.

ABOVE Mud-walled cells built and provisioned by a female red mason bee (*Osmia bicornis*) inside the wall of a garden shed.

BELOW A few solitary bees are primitively eusocial. Sharp-collared furrow bees (*Lasioglossum malachurum*) live several to a nest with a dominant 'queen' female.

Females and males emerge in spring, after having spent the winter inside the nests built by their mother the previous year. Like bumblebees, different species emerge at different times. The earliest solitary bee you are likely to notice in your garden is the charming and delightfully named hairy-footed flower bee (*Anthophora plumipes*), which in southern UK sometimes flies from late February. By contrast, the wool carder bee (*Anthidium manicatum*) is not usually seen until late May. Males often emerge first, giving them time to forage, familiarise themselves with their surroundings and become sexually mature, ready for the females as soon as they emerge a little later. Some males can't wait that long and will burrow frantically into nests and drag females out to mate with them. Others gather in large groups at the nest entrance, fighting off rivals and mobbing emerging females in a seething mass called a mating ball. The males of other species show more patience, flying circuits around their territory until they find females to partner with in a slightly more dignified way. After mating, males have no further role and live out their days flying and foraging – often for only a matter of a few weeks.

Once mated, females seek out suitable locations for their nests. These are built in varied places using different techniques as explained below, but once completed they are generally used in much the same way. Females collect pollen and nectar to provision the nest they have built. Most species are polylectic, meaning they will forage on a wide range of suitable flowers. Some are oligolectic and rely on a very limited selection of flowers from just a single genus or group. Three UK bees are monolectic and rely probably solely on a single species – without that one flower they cannot survive. Pollen is gathered on stiff hairs on and around the legs, or in some species using a brush of hairs on the underside of the abdomen. Some, such as the tiny, yellow-faced bees of the genus *Hylaeus*, swallow the pollen and regurgitate it when they get back to the nest. Once a brood chamber has been provisioned with a suitable quantity of pollen and nectar, a single egg is laid and the chamber is sealed. The female bee then builds

ABOVE Male wool carder bee (*Anthidium manicatum*) mounting a female.

ABOVE Female solitary bees have various ways of collecting pollen. This patchwork leafcutter bee (*Megachile centuncularis*) has fully loaded pollen brushes (scopae) on the underside of her abdomen.

another chamber, continuing in this way until she has filled the nest space or has no more eggs to lay.

After hatching, the larvae consume the provisions around them. They grow rapidly until they are ready to pupate and begin metamorphosis, their body parts reforming and rearranging themselves to produce an adult bee. Some go through most of this process straight away, to produce almost fully formed bees which then remain in the same state through winter before emerging the following spring. Others develop only partially before entering a dormant winter state called diapause, development resuming again before the bees emerge the following year. Most solitary bees produce only one generation annually and are described as univoltine. Bivoltine species on the other hand produce two generations annually; the bees that emerge in the early part of the season make nests, lay eggs and die before their young emerge and lay their own eggs, which will become the next year's early bees.

As with bumblebees, some solitary bees are cuckoos – brood parasites that lay their eggs in the nest of another species. Their larvae kill the egg or larva of their host before consuming their store of pollen and nectar. Almost 70 species of UK solitary bees behave in this way.

Identifying solitary bees

So far, this book has dealt with just two genera of bees: *Apis* (the honey bees) and *Bombus* (the bumblebees). Solitary bees in the UK comprise a further 26 genera, some of which contain dozens of species. There are about 256 species of solitary bee in the UK – the exact number is somewhat fluid as new species occasionally spread from mainland Europe, existing species go extinct, and some thought to be extinct are sometimes rediscovered. For example, in 2021 the Maidstone mining bee (*Andrena polita*) was recorded in Kent, having not been seen since the 1930s. Twenty-five species of our solitary bees have been recorded as now extinct in the UK.

Solitary bees are highly varied in behaviour and appearance; some are large and fluffy, resembling a small bumblebee, some are only about the size of an ant, while others look remarkably like wasps. Some solitary bees have unique features and are easy to recognise, while many differ from one another in ways so subtle that only a microscope (and possibly a degree in entomology) will help you to work out exactly what you are looking at. Males and females often look very different from one another, adding another layer of complexity. Females are generally larger and more attractively coloured and patterned than males – though not always. Remembering the scientific names of solitary bees can be a real challenge too, even for the dedicated bee watcher. Fortunately, most of the solitary bees likely to be found in gardens have common names that are relatively easy to remember. These often refer to a bee's nesting habit and sometimes some other distinguishing feature. For example, the common name of the not very common and otherwise tongue-twisting *Andrena nitidiuscula* is carrot mining bee, so called because it commonly forages on wild carrot flowers and mines tunnels in which to lay its eggs.

The distribution of solitary bees depends on the availability of suitable nesting sites and forage plants. This means that many species are found only in small or fragmented areas of the country. Some species are rare even in their strongholds, while others can be locally very common. This means that of the more than 250 species in the UK, only a few dozen are likely to be seen regularly near where you live. However, one garden in Surrey has been recorded as hosting 133 species of bees, and a small garden in the built-up centre of Bristol has recorded more than 50 species. Consult your local wildlife trust or a good field guide to find out what bees you can expect to see in your area, and which to support when planting and creating garden habitat.

To help aid identification as well as an understanding of their lifestyles, solitary bees can be divided broadly into two groups according to the way in which they build their nests – ground-nesting bees and cavity-nesting (sometimes called aerial-nesting) bees. The following is a description of some of the species in each category that are more commonly found in UK gardens, parks and hedgerows.

Ground-nesting bees

These are species in which the females build their nests by underground tunnelling. The location chosen varies according to the species, from flat, clay soil to sandy banks, cliff faces, the root-plates of fallen trees and even within the soft mortar of brick walls. The tunnels can be anything from a few centimetres to several tens of centimetres in length, and may consist of a single tunnel or one that branches in different ways to form small chambers. These will be provisioned with food before an egg is laid and the chamber is sealed. The tell-tale sign of a ground-nesting bee nest is a small volcano of excavated sand or soil with a hole at its centre. If you keep an eye out in spring and summer it is surprising how often you will spot these in lawns, hedge banks and even in the middle of well-worn pathways. Take a minute or two to watch and you might see a bee emerging, or possibly a pollen-laden female arriving home. Sometimes there will be a single hole, sometimes an aggregation with dozens or even hundreds of holes in a small area. Some species fill in and hide their hole each time they leave, while others build decoy holes to attract predators away from their nest.

The biggest group of ground-besting bees with this behaviour in the UK are those in the genus *Andrena*, which comprises around 1,500 species worldwide. In the UK there are almost 70 species of *Andrena*, varying greatly in appearance, and ranging in size from as little as 5mm to as much as 17mm in length. The common factor is that they all build tunnels when nesting, earning them the common name of mining bees.

Andrena mining bees mostly nest in light or sandy soils, usually in flat ground but also in vertical surfaces like cliff or quarry faces. They sometimes nest in aggregations, with many bees building tunnels within a small area. To help to prevent their tunnels from collapsing or becoming damp, or the stored pollen from going mouldy, females secrete a waxy substance which soaks into the walls to form a damp-proof lining.

Tawny mining bee (*Andrena fulva*)

This relatively common mining bee might alternatively be called the flying fox because of the glorious, foxy-orange furry appearance of the females, which are often spotted basking on leaves in early spring. The strikingly coloured fur is on the top of the thorax and abdomen, but the head, legs and undersides are a plush velvety black. It can be seen foraging on flowering shrubs and trees like hawthorn, blackthorn, maple and apples, as well as on umbellifers, buttercups and dandelions. Frequently found in urban areas, the tawny mining bee commonly nests, sometimes in aggregations, in lawns, close-cropped grassland and garden flowerbeds. The male is smaller, less hairy and an overall sandy colour, with a banded abdomen and white face.

Female

Male

Ashy mining bee (*Andrena cineraria*)

Female

Male

One of our most distinctive species, the female ashy mining bee has a shiny, blue-black abdomen and wide bands of ashy-white hair on the thorax and front of the face. This coloration has led it to being nicknamed the 'panda bee', but 'bandit bee' might be more appropriate because, combined with its monochrome appearance, the black stripe over the eyes is reminiscent of the masks worn by bandits in old movies. Males are smaller and fluffier, with an overall grey appearance. They appear in early spring and nest, often in large aggregations, in short grass such as that nibbled by sheep or rabbits, on south-facing slopes, on country tracks and pathways and sometimes lawns. They visit a wide range of spring flowers.

Orange-tailed mining bee (*Andrena haemorrhoa*)

Typically seen from about late March until mid-July, this pretty and widespread bee is slightly smaller than a honey bee. It has a rich-red thorax, though not as bright as that of the tawny mining bee's, with shorter and less shaggy hair. The abdomen is dark grey with a few white hairs and a neat orange tip to the tail. Nests can be found in a wide range of places, including urban gardens and intensively farmed land, and even in heavy clay. It forages on a wide range of spring and summer flowers. This species is sometimes also called the early mining bee.

Male

Female

Chocolate mining bee (*Andrena scotica*)

With an overall chocolatey-brown hue, this is one of the UK's most common and widespread mining bees. Females are about the size of honey bees and look superficially similar, but they have an overall darker, chocolatey-brown appearance and longer antennae. Males are smaller, slimmer and a less rich brown. The best time to see them is when blackthorn and hawthorn are in flower, though they visit a wide range of other spring and summer flowers. They are often found in urban gardens, nesting singly or in loose aggregations, usually in less open areas such as hedge banks, or among longer grass and leaf litter. They are widely parasitised by *Nomada* cuckoo bees and beeflies (see page 206).

Female

Male

Common mini-miner (*Andrena minutula*)

One of a group of small-sized *Andrena* species known mini-miners, this bee measures only about 7mm in length. It is largely black with faint fringes of light hair on the abdomen, and the wings have a metallic sheen. It visits a wide range of flowers between late spring and late summer. It is bivoltine (having two generations in a year), with the first generation emerging in mid-March, and the second in about mid-June. This is a good example of a small solitary bee that many people might barely notice, or even dismiss as a fly. Looking for small bees such as this really opens your eyes to how different in size, appearance and behaviour solitary bees can be compared with honey bees and bumblebees.

Female

Ivy bee (*Colletes hederae*)

There are nine species of the genus *Colletes* in the UK and an estimated 700 worldwide. Known commonly as plasterer bees, members of this genus paint the inside of their tunnels with a liquid polymer that dries into a cellophane-like material. The liquid is produced by a gland in the abdomen and collected and plastered onto surfaces with the tongue. Each sealed brood cell is like a waterproof plastic bag that holds a liquid mixture of pollen and nectar, as well as the egg. Being watertight, this can keep the young alive even through winter floods. The ivy bee (*C. hederae*) was not recorded in mainland UK until 2001, having spread from mainland Europe. It has spread rapidly and is now found commonly as far north as the Scottish borders. This charming fawn and black striped bee emerges in early autumn and

Male

feeds almost exclusively on ivy flowers. It sometimes visits other flowers if ivy is not yet available. It often nests in large, highly populated aggregations on south-facing sandy ground with patchy grass, often in lawns. Males patrol nest areas and pounce, sometimes in their hundreds, on emerging females, creating a mass of bees called a mating ball.

Davies' colletes (*Colletes daviesanus*)

This species looks remarkably similar to the ivy bee but is a little smaller and can be seen throughout the summer, generally disappearing around the time the ivy bee appears. It is often found in gardens, foraging on a wide range of flowers. It can nest in large aggregations on sparsely vegetated dry soils, especially on dry slopes, but also sometimes in walls with soft mortar.

Female

Common furrow bee (*Lasioglossum calceatum*)

The genus *Lasioglossum* comprises around 1,700 species worldwide, making it one of the largest genera in the animal kingdom. Along with members of the smaller genus *Halictus* (see opposite) the 33 *Lassioglosum* species found in the UK are commonly called furrow bees because the females of both genera have a distinct hair-lined 'furrow' on the last overlapping segment (tergite) of the abdomen (the tip of the tail). They are also sometimes called sweat bees, because in warm climates some members of this genus are attracted to perspiration.

Furrow bees are very diverse, ranging from the tiny *L. minutissimum* at 5mm in length, to *L. xanthopus*, which is about the size of a honey bee. Some species in the genus can be both solitary and eusocial, meaning they can have a queen who produces workers that go on to raise new queens and males. Only the queens survive over winter.

As its name suggests, the common furrow bee (*L. calceatum*) is the most widespread and frequently seen species. The female is very attractive, with a glossy, dark brown abdomen bearing bands of brown and cream hairs. Males are slender and dark in colour, with light bands on the first half of their abdomen and white legs. Some males have a very attractive dark red abdomen. The males sometimes roost in clusters, rather charmingly spending the night huddled together inside flowers or under leaves. They forage on a wide variety of flowers, particularly thistles and knapweeds, and are found in a range of

Female

Male

habitats, including urban gardens. Nest tunnels are built in light soils, sometimes in small aggregations. They are often one of the last solitary bees to disappear before winter, sometimes still flying in October.

BELOW Male common furrow bees preparing to spend the night together on knapweed.

Orange-legged furrow bee (*Halictus rubicundus*)

A widespread and commonly seen solitary bee. Females are about 10mm in length. They have gingery hair on the face and thorax, and a smart, glossy black abdomen with bands of white hair. The most striking feature of this bee are the creamy-orange back legs, with long hairs for collecting pollen. Males are slimmer and less boldly coloured, with pale yellow legs. This species is found in a variety of habitats on a wide range of flowers, nesting in light soils, sometimes in large aggregations. Interestingly, it is generally eusocial (raising queens and working as a group to raise young) in the south of the UK, but solitary in the north where conditions are cooler and the summers are shorter. It has been extensively studied for this social plasticity, and has contributed to our understanding of the evolution of social behaviour.

Female

RIGHT Notice the small furrow in the last segment of the abdomen, from which the various species of furrow bee get their name.

Common yellow-faced bee (*Hylaeus communis*)

There are about 500 species of *Hylaeus* worldwide, with 12 species found in the UK. They are small, slender and predominantly black and hairless. Their most distinctive features are the yellow facial markings from which they get their common names. They are plasterer bees, lining their nest cavities with a plastic-like material, similar to the *Colletes* species mentioned earlier. Uniquely, pollen is swallowed and transported to the nest where it is regurgitated along with nectar to make a liquid food that is sealed into cells along with the eggs.

The common yellow-faced bee is found throughout the UK, but less so in the north. Females measure about 6mm in length and have triangular yellow markings on the face and yellow markings on the legs and thorax, but are otherwise completely black. Males are slightly smaller and have more yellow on their faces. Found in a wide range of habitats and on a variety of flowers, usually from late May until mid-September, they are opportunistic nesters and will use not only tunnels in soil but also hollow plant stems, holes in rotten wood and timber, and often the smaller-sized holes in garden bee nesting boxes.

Hairy-footed flower bee (*Anthophora plumipes*)

Male

Female

ABOVE Female foraging on perennial wallflower.

RIGHT Female approaching nest hole made in crumbling mortar.

Commonly known as flower bees, there are about 450 species of *Anthophora* worldwide, with just five species in the UK. One species, the potter flower bee (*A. retusa*) was once quite common but has declined hugely and is now found only in a handful of places. Flower bees are chunky, furry, highly charismatic bees with prominent eyes. They are easy to spot and are often confused with small bumblebees.

The hairy-footed flower bee (*A. plumipes*) is usually the first solitary bee in the UK to emerge each spring, with the gingery-brown males – often confused with common carder bumblebees – typically appearing in early March. The black and dark grey-striped females emerge about two weeks later. They visit a wide variety of flowers, but lungworts, deadnettles and comfreys are a favourite. They establish circuits between patches of flowers and zip energetically between them, hovering briefly in front of flowers with their tongues extended in a hummingbird-like manner before zooming in to collect nectar and pollen. The males, who have extravagantly feathery middle legs, court females very enthusiastically, chasing them and often knocking them off flowers when they are feeding. Nesting can be take place in large aggregations in soil banks, the faces of quarries or sides of rabbit burrows. They often make holes in the soft mortar of old brick or stone walls. They usually finish nesting and disappear by about mid-June. These bees are real characters and once you have started to notice them you will spot them regularly as they dash around patches of flowers on sunny late-spring and early-summer days.

Cavity-nesting or aerial-nesting bees

Unlike mining bees, which build their own nesting tunnels, most of these bees take advantage of pre-existing holes, gaps and crevices, often in the vertical surfaces of trees and walls, to build their nests. Chosen nest sites are modified as needed, using a variety of methods and materials to create the chambers in which provisions and eggs are placed. This group includes the species most likely to occupy artificial bee nesting boxes.

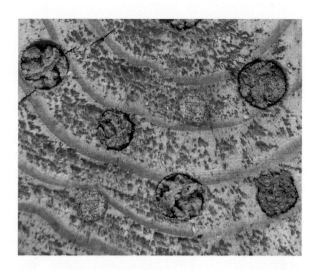

RIGHT A nesting block used by solitary bees. The smaller holes, filled with plant material, have been used by blue mason bees (*Osmia caerulescens*), the larger holes, filled with mud, by red mason bees (*O. bicornis*).

Red mason bee (*Osmia bicornis – previously Osmia rufa*)

There are more than 300 species of mason bees, genus *Osmia*, in the world, with 12 found in the UK. They are medium to large bees and are often colourful. They build nests in existing cavities and have large jaws to help with gathering materials and construction. Their ability to build walls and tunnels using gathered materials earns this group the common name of mason bees. They are important crop pollinators, and in the USA are bred on a commercial scale for this purpose.

Female

The red mason bee is perhaps the most widespread and commonly seen UK solitary bee. It is also the species most likely to use bee nesting boxes. Females are about 11mm long, furry all over with a black head, gingery-brown thorax and ginger-orange abdomen. Males, which are smaller, slimmer and duller, with noticeably longer antennae, usually emerge in late March, followed by females about two weeks later. They are found especially in gardens, churchyards, parks and urban areas generally. They visit a wide range of spring flowers and will usually be seen on apple and pear blossom. Nests are made in a variety of pre-existing holes, including garden canes, hollow stems, crevices in old walls, beetle holes and bee nesting boxes. The females collect mud from around ponds, puddles and watered garden beds, carrying it in their mandibles and using it to line cavities and build partition walls between brood cells. They have a pair of stubby horns on their face which are used for sculpting mud in the nest.

Male

Blue mason bee (*Osmia caerulescens*)

These small, but chunky-looking bees with box-like heads and large eyes are mostly black, but with metallic bluish highlights. Females have white bands on their abdomen. Males are smaller and with gingery-yellow hairs. They are frequently seen in parks and gardens and though less common and slightly smaller than *O. bicornis* will often nest in the same artificial nest boxes, usually using slightly narrower nesting tubes. The nests can be told apart by the material used to seal them; the red mason bee uses mud, whereas the blue mason bee uses finely chewed leaves and sometimes petals.

ABOVE Female on yellow monkswort.

LEFT Female showing blueish highlights.

Red-tailed mason bee (*Osmia bicolor*)

The *bicolor* part of this bee's scientific name describes the female of the species, which is black at the front and red at the back. Males are smaller and a rather drab brown colour. It is locally common on unimproved grassland in chalk and limestone areas of southern UK, particularly south-facing slopes, with populations extending only as far north as the Midlands. This species is unlikely to be seen in your garden, but it is worth taking downland walks in early summer to look for it. Apart from its striking appearance, this bee's main attraction is its charming nesting method. It builds a brood nest from chewed-up leaves plastered inside an empty snail shell. After eggs are laid, the shells are plugged with mud, pebbles or crushed snail shell and then hidden under a wigwam of grass, pine needles or leaves.

Female

Leafcutter bees (*Megachile* spp.)

There are 1,500 members of the *Megachile* genus worldwide, with seven species recorded in the UK. They are commonly known as leafcutters because they fashion their nests out of pieces of carefully cut out plant leaf. To do this they have large, sharp jaws that can be used in a scissor-like way. To cut a section of leaf, the bee secures itself to the edge and pivots in a complete circle, snipping as it turns like the bee equivalent of a tin opener. The leaf is then rolled into a tube and transported to the nest site, where its edges are chewed to form a sappy glue that is used to seal the tube and stick it to others. Walls and plugs are also made from leaf pieces cut to just the right size. Leafcutters are chunky-looking bees with noticeably ribbed-looking abdomens. The underside of the abdomen is covered with orange or yellow pollen-collecting hairs, which are often easy to spot as the bee tends to raise the abdomen repeatedly while foraging.

Willughby's leafcutter bee (*M. willughbiella*) is the most common leafcutter in the UK and is a frequent visitor to gardens, where it forages on a wide range of flowers. Females are large and furry with a dark brown and cream striped abdomen which has bright orange hair on the underside. Males are similar but an overall chestnut-brown colour with dainty-looking white front feet. It flies from around late May until late August. This is the leafcutter species most likely to use bee nest boxes, usually taking advantage of the larger-sized holes. It also nests in rotten wood, between fence panels, in gaps around doors and windows and even inside unused garden hose pipes. Even if you don't see the bees, you may notice circles cut out of leaves. Roses are a favourite, but they will use lilac, wisteria and grapevine, among others. The patchwork leafcutter bee (*M. centuncularis*) is the other most commonly seen UK leafcutter.

ABOVE Female patchwork leafcutter sealing her nest.

ABOVE Individually wrapped leafcutter cells inside a bee nesting box.

ABOVE Female cutting out pieces of grapevine leaf. Each section takes about 5 seconds to remove.

ABOVE Male Willughby's leafcutter with hairy, white front feet.

Wool carder bee (*Anthidium manicatum*)

There are several hundred species of wool carder bee, but only one is found in the UK. This magnificent bee is unlike any other and is relatively easy to find and to identify. Males usually emerge in late May and set up territories around favoured species of plant, particularly woundworts and black horehound. In gardens they are closely associated with lamb's ear, great mullein and yarrow. They fiercely guard these territories, chasing away and even headbutting any other flying insects, including much larger bumblebees, and sometimes even hovering at eye level to challenge humans. The tip of the abdomen is sharply spined and can be used in mid-air to damage or kill intruders that don't get the message. A mated female collects woolly nesting material from the leaves and stems of hairy plants, shaving it off and then carding it into a ball which she carries tucked between her legs to the nesting site. This could be

Female

any suitable cavity, such as a hollow stem, beetle hole, bee nesting box (occasionally) or even inside garden furniture. The wool is fashioned into knitted pockets, within which provisions and eggs are placed.

This is a rare example of a species in which the males are bigger and more colourful than the females, being large, chunky and dark grey-brown with lighter grey fringes of hair. The abdomen appears ribbed, and has spines at the tail end and distinctive bright yellow spots along the sides. Females are smaller, less robust-looking, darker and with fewer hairs, and have a similar arrangement of yellow spots.

Male

Cuckoo solitary bees

Incredibly, there are 67 species of cuckoo solitary bees in the UK. Each is a brood parasite of just one or a few host species. Their usual modus operandi is to sneak into a nest while the homeowner is away collecting pollen or nesting materials, and to lay their own eggs. Their larvae then eat the host's offspring and the store of food.

The most widespread and easily spotted cuckoos belong to the genus *Nomada*, commonly known as nomad bees. Many of these are yellow, striped and strikingly waspish in appearance. They can often be seen hanging around solitary bee nest holes, looking decidedly shifty and waiting for their opportunity to invade. It is sometimes their presence that gives away the location of solitary bee nests. There are more than 30 species of nomad bees in the UK, many of which specialise in parasitising species of *Osmia* (mason bees).

Another species of the genus is *Sphecodes*, commonly known as blood bees because most of them have a translucent-looking blood-red abdomen. The 17 species of *Sphecodes* specialise in invading nests of the genera

ABOVE Dull-vented sharp-tailed bee (*Coelioxys elongata*) investigating the nest of a leafcutter bee.

BELOW The wasp-like cuckoo bee *Nomada flava*.

Lasioglossum, *Halictus* and *Andrena*, all ground-nesting bees. Members of the genus *Coelioxys* are known as sharp-tail bees. Their abdomen has a pointed tip, which they use to pierce brood cells and inject their eggs. They parasitise species of *Megachile* (leafcutters) and *Anthophora* (flower bees).

Nest boxes for solitary bees

Bird nesting boxes have always been a mainstay of the wildlife-friendly garden, and there can be great pleasure in watching them being chosen and used by birds – and later in seeing the successful fledging of their chicks. More recently, bee nesting boxes or 'bee hotels' have become popular, even fashionable. But as with bird boxes, there can be a big difference between what well-intentioned humans think looks homely and what prospective inhabitants will find appealing; dainty roofs and pastel colours count for nothing when solitary bees are looking for somewhere to nest.

Although there are some good-quality off-the-shelf bee nesting boxes, many have basic design flaws, can be dangerous for bees and rarely come with helpful instructions. The requirements in fact are quite straightforward, and it is not difficult to build your own, even with only rudimentary DIY skills.

However, most species of solitary bee won't use any kind of nesting box. Ground-nesting bees use tunnels in undisturbed lawns, banks or walls. It is quite possible to create potential nesting sites like these in your garden – though it's not as convenient or instant as nailing a nesting box to a tree, and their adoption is a bit more sporadic. Bumblebees have different nesting requirements altogether and, although artificial nesting boxes of various designs can be made or purchased, uptake tends to be rather limited.

Although it is nice to see bees nesting in your garden, other than for a few species it's hard to tell if such additional opportunities significantly improve overall populations. One undoubted advantage is the ability to provide nesting sites that are protected and well managed. Wild bee nests in parks, roadside verges, hedgerows and possibly even in your neighbour's garden are vulnerable to obsessive mowing, hedge-trimming and general over-tidiness. Nesting tunnels can be overturned by shovel or plough, or be poisoned by agrochemicals. Bad weather and predators can also take their toll. Such threats can be limited if nesting sites are provided in the controlled environment of the bee-friendly garden, and if some simple maintenance is undertaken. In such circumstances, local populations can thrive, with some species – which can be quite loyal to the place of their birth – often nesting in the same site year after year, their numbers in the immediate area growing noticeably.

Providing nest sites for bees and seeing them make good use of them is tremendously rewarding and satisfying – arguably even more so than watching nesting birds. Watching bees as they gather materials, build nests and bring back pollen and nectar provides insights into their fascinating behaviour and gives you

ABOVE A good example of a bad nest box; the tubes are far too wide and short to be of any use.

ABOVE Homemade nest boxes are easy to make. This one, being used by the common yellow-faced bee (*Hylaeus communis*), uses reeds gathered from a roadside ditch.

a connection with the wildlife on your doorstep. It can also be a thrilling educational experience for children, and a wonderful introduction to the world of what schools today call 'mini-beasts'.

The requirements for cavity-nesting solitary bees are quite straightforward. They look for a hole of about the right diameter, which leads to a tunnel of a useful length with no obstructions and with a sealed-off far end. It should be sufficiently sheltered from the elements to ensure that nest materials and the developing larva within remain dry. Some commercially available boxes manage to fulfil these basic requirements, and if you wish to buy one it is worth looking at those offered by suppliers with a specialist knowledge of wildlife. However, there are lots of ways to build your own, most of which don't require sophisticated DIY skills, cost next to nothing and will be every bit as successful as anything you might buy. Be aware that, as with bird boxes, not all bee nesting boxes will be occupied. In some areas the bees most likely to use nesting boxes are uncommon, or they just might not find your box or might prefer other established nesting sites nearby. Be persistent and reposition unused, clean boxes each year; you might find a sweet spot and be rewarded with nesting bees who should return annually from then on. It is quite likely that various species of solitary wasp will nest in your boxes too. These are perfectly harmless to you and in most cases don't bother the bees. They are every bit as vital to the ecosystem as are bees, and are worth welcoming.

Building a box

Unlike bird boxes, which need to have the right dimensions for different species, bee nesting boxes can be almost any size and shape you like. Avoid very large boxes with hundreds of cavities, as these are hard to manage and possibly put the bees at increased risk from disease, predators and parasites. If you want to provide lots of nesting opportunities, it is better to make several smaller boxes. These can offer varying nesting conditions and be dotted around the garden, increasing the chances of successful occupancy.

First, you need a box. This doesn't have to be homemade; if you have access to some old crates or containers of about the right size, they will do handsomely. If making your own boxes, you can be as

ABOVE A homemade log nest block, used by several species.

ABOVE Simple wooden boxes filled with drilled wooden blocks and cardboard nesting tubes.

ABOVE Nesting block being used by large-headed resin bee (*Heriades truncorum*).

creative as you like in terms of shape and design as long as you follow the basic principles. Use exterior-grade sheet materials or planks of untreated timber. Old floorboards are ideal and will be all the more attractive. For a basic design, screw or nail them together to make a box that is around 20cm square and 20cm deep. One side will be the roof when the box is hung outside; try to make this piece overhang by at least 5cm, and if you can make it pitched to give additional protection from rain, even better. The box does not necessarily need a back wall, though this depends on where it will be placed and what kind of nesting materials you plan to use. It is fine to paint the outside surfaces of the box with water-based paint or varnish.

Nesting materials

For most species of cavity-nesting bee, holes in wood are ideal. Use blocks of wood or logs that will fit into your nest box and that present a reasonably flat front surface. They should be slightly shorter than the depth of the box so that they will be a little recessed when installed – giving more shelter from the elements. Whatever you use, try to make sure it is well seasoned and close-grained, so that it is unlikely to split. Avoid anything that has been treated with wood preserver.

Drill holes in the wood, spacing them about 3–5cm apart. The diameter of the holes will determine which species of bees might nest in them. Use sizes ranging 2mm to 10mm. The smallest will be used by tiny species such as the yellow-faced bees (*Hylaeus* spp.), scissor bees (*Chelostoma* spp.) and resin bees (*Heriades* spp.). Holes of 6mm–8mm are typically used by mason bees (*Osmia* spp.), and 8–10mm holes might be used by leafcutters (*Megachile* spp.). In general, 8mm holes seem to get most occupancy, so provide more that are that size.

Larger-diameter holes (6mm and above) should be drilled at least 15cm deep – even up to double this depth – and not go all the way through the material, giving the tunnels a closed end. Small-diameter holes can be shallower, about 8cm deep. A set of long-shank drill bits is a worthwhile investment, and a pillar drill makes the job much quicker. Drill each hole several times to remove sawdust and splinters, and make sure the entrance of the hole is clean and smooth; this can be done with a counter-sink drill bit or by running sandpaper around the rim.

Place the drilled pieces into your nest box, ready for hanging in the garden. If there are gaps, you can offer additional nesting opportunities by filling them with some of the tubular nesting materials described below.

Free-standing nesting blocks

Large blocks of wood or logs don't even need to go into a box. Simply drill holes in them and mount them directly on a wall or tree. Old roof joists are ideal for this and are often to be found in builder's skips. Old fence posts are good too, but avoid anything soaked in wood preserver. A piece of wood or an old roof tile fixed over the top will give some shelter from rain.

If you have a dead tree in your garden, leave at least the trunk standing and drill holes in the sunniest side.

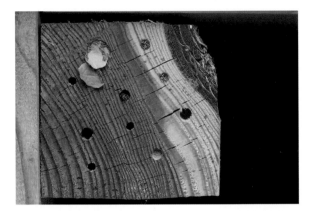

ABOVE The end of a shed roofing joist drilled with bee nesting holes. Several species are using it.

ABOVE Nesting holes drilled decoratively in a dead tree.

It will make a spectacular feature when bees begin to nest in it. Beetles may bore holes in old trunks, and in time these will attract bee species that nest in very small tunnels. Large dead branches and logs can also be used. You can be as creative as you like with the pattern of the holes you drill in them. Standing in a border and slowly bleaching and decaying, they can be as attractive as any garden sculpture and far more beneficial. If you wander around the garden with a cordless drill, it is surprising how many things you can find that might be given a few experimental holes – it can become a bit addictive, as can checking them later for occupancy.

Nesting tubes

Rather than drilling holes, individual nesting tubes can be used. These can be inserted in nesting boxes, into lengths of drainage pipe, or pushed into plastic bottles with the neck removed. Placing small clusters of them in multiple sites around the garden is a good way to find out which positions are favoured by bees.

Cardboard and paper tubes designed specifically for solitary bees are available to buy online. Some come with stoppers at one end – if they don't, make sure they are pushed tightly against the back wall of the container you put them in, or plug them with small pebbles, mud or modelling clay. Cardboard tubes easily get damp, so keep them sheltered from rain and bring them inside over winter (see below). Some are designed to unravel when dampened, allowing cocoons to be removed for overwinter storage. If you buy this type, they usually come with instructions. Paper tubes are easy to make and surprisingly effective. Simply wrap thick cartridge paper around a piece of wooden dowel of the correct diameter and secure it with a small piece of sticky tape. Slide the tube off the dowel and staple or tape one end to close it. The resulting tubes can be used on their own or as liners in more durable cardboard, bamboo or plastic tubes, allowing the paper tube and any cocoons to be easily removed before winter.

There are plenty of options for tubes made of natural, sustainable materials. Bamboo canes, often used in commercially made bee nesting boxes, can be very successful. Buy long lengths from garden centres and cut them to size. Make sure any knots are at the far end and that the tubes are otherwise clear. Cut with a fine-toothed saw to avoid sharp splinters, and smooth down the ends with sandpaper.

ABOVE Female red mason bee nesting in a bamboo tube.

ABOVE Sealed nesting tubes in a length of drain pipe.

ABOVE Cardboard nesting tube unwrapped to remove red mason bee cocoons.

Gardens and hedgerows are full of hollow-stemmed plants that can be used in bee nesting boxes. In late summer as they die back and dry out, collect stems from brambles, teasels, raspberries, elder, umbelliferous plants such as cow parsley and wild carrot, and various reeds and grasses – anything with an internal diameter between 2–10mm. Reed stems can be particularly successful in nest boxes. Keep the materials dry until spring, and then bundle them together and trim the ends so that they are even and splinter-free. They work well clustered in empty tin cans or plastic bottles, and can be used to fill gaps between wooden blocks in nesting boxes.

Groovy nesting boxes

If you have a router or table saw, nest holes can be cut into planks or blocks of wood. The tunnels produced in this way will be open-sided, so they must be pushed up against another surface to replace the missing wall. One method is to use square-profile blocks of wood and then stack them next to each other inside a nesting box so that each block provides a solid wall for the tunnels in the adjacent block, or is pushed up against the wall of the box itself. Another option is to fix transparent Perspex to the side of the blocks, to provide the missing tunnel wall. After having been used by bees, the blocks can be removed and the nest contents viewed through the Perspex. Yet another alternative, now quite common in commercially available nest boxes, is to use wooden drawers, each about 3cm thick (i.e. a sawn-up plank of wood), with tunnels cut into the top surface. When stacked vertically, the underside of each drawer provides the ceiling for the tunnels in the drawer below.

BELOW Red mason bee cells in a grooved nest box tray.

ABOVE Grooved nesting box, with bottom right-hand cavity used by a leafcutter bee.

When nesting is complete, the drawers can slide out for inspection or removal of the cocoons for safe storage over winter.

Siting, predators and maintenance

Where you place a nesting box is key to its success. Trees, walls, sheds or fence posts can all work well. Solitary bees like a sunny nesting spot, so fix boxes so that they face south or south-east. Choose sheltered spots relatively free of overhanging leaves or branches, so that the boxes get maximum exposure to the sun and avoid driving rain, which can make nests and nesting materials damp. Place them at least a metre off the ground – at about your eye height is ideal. Fix them securely so that they don't swing or sway in the wind. Put them in place by about mid-spring, and then look out for bees coming and going, carrying nesting materials or with bright yellow pollen loads. It is fascinating and quite safe to get very close to the nests to watch the bees' behaviour as they seal the ends of their tunnels with mud, leaves and other materials, depending on the species.

Red mason bees require soft mud to line and seal their nests. They can normally find this at the edges of puddles or ponds, but if it is dry, or if you just want to watch them at work, provide a source of damp muddy clay near the nesting box. This can often be achieved by digging a hole in a border or vegetable patch. Water it regularly to keep it moist enough for the bees to use.

Bee nest boxes can be left outside all winter, and in most circumstances young bees developing inside should survive and emerge the following year. However, you can improve their chances by taking a few precautions. Some birds learn to peck at the tubes to get at the bees; great tits might get at the first one or two in a tunnel, but woodpeckers can use their long bills and even longer tongues to reach further in. If you notice birds taking an interest, cover the front with a piece of wire mesh, securing it at a distance so that they can't reach the nests. Alternatively, turn the nesting blocks or tubes around to face the back wall of the box – turn them back again in late spring. Mice can sometimes be a problem. If you notice nesting materials starting to get nibbled, remove them and store as recommended below.

Solitary bees and their nests are vulnerable to various parasites, some of which can kill the developing larvae. These of course have as much right to exist as their prey, and their presence is probably a sign of a healthy bee population. However, nesting boxes can cause bees to breed in more congested conditions than they might do naturally, and they can become a target for greater numbers of predators and parasites and a repository for potentially harmful pathogens.

Having a few parasitic wasps lay their eggs in one or two nest tunnels perhaps should not be begrudged. Some, such as the gloriously named *Gasteruption jaculator* – commonly called the javelin wasp – are quite beautiful in a sinister kind of way.

One common parasite is *Cacoxenus indagator*, a tiny, red-eyed fly that attacks only the nests of mason bees – the most frequent inhabitants of nesting boxes. The adult fly sneaks into nesting tubes and lays its eggs on the pollen stored there. After hatching, the fly's larvae eat the pollen, starving the bee larva which itself is then eaten. After pupating, newly emerged but trapped flies find small cracks and fissures in the dividing walls of the nest. Burying themselves in these they inflate their bladder-like head to about four times its original size, pushing the dried mud to one side or making dividing walls crumble and fall away. This amazing escape technique has earned the species its common name of the Houdini fly. Populations can build significantly over several years if allowed to continue breeding uninterrupted in a bee nesting box.

Another bothersome creature is the tiny pollen mite, *Chaetodactylus osmiae*. These minute spider-relatives

ABOVE Female red mason bee using a nest box placed in full sun.

ABOVE Nesting tubes in a wooden shelter on a sunny wall. Wire mesh protects the nests from birds.

enter mason bee nests when female bees are provisioning them with pollen. They eat the pollen and multiply rapidly, often starving the bee larva, which is eaten too. Thousands of mites can accumulate in a cell, producing nymphs that will develop into adults the following spring. Any surviving bees will have to push their way out through a mass of mites which then cling onto their body for transportation, hitching a lift to flowers where they await a fresh victim, or jumping onto another bee during mating. Bee nesting boxes that are unmanaged for several years can acquire huge populations of these mites, leading to a collapse in the bee population.

If your nest boxes attract leafcutter bees you may see their tunnels being used by sharp-tailed bees (*Coelioxys* spp.), whose larvae will consume the stores and possibly

ABOVE *Cacoxenus indagator*, a tiny parasitic fly often seen around nest boxes.

ABOVE Red mason bee nest invaded by *Cacoxenus indagator*. The fly's larvae are surrounded by their droppings.

ABOVE *Gasteruption jaculator* scouting a nest.

even the eggs and larvae of the leafcutters. There is not much you can do about this, and anyway it is only a matter of one native bee species surviving at the expense of another.

You could be relaxed about the danger of various predators and parasites and let nature take its course, but if your reason for having nesting boxes is to benefit local bee populations then some precautions are perhaps worthwhile. One method, which can be great fun if done with children, is to harvest bee cocoons and clean out nest boxes at the end of the nesting season. Cocoons harvested in this way will be from either mason or leafcutter bees, and this is only possible with nesting boxes that allow access to the cocoons. When nesting has finished (in the UK, typically in July, after leafcutters have nested), access the cocoons by unwrapping cardboard tubes, splitting open canes or reeds or exposing tunnels in wooden blocks. Cardboard tubes designed to unravel when wet should only be dampened if you know they have mason bee cocoons in them (mud entrance plugs) and not leafcutter cocoons (leaf plugs), which will collapse when damp. Bamboo and reed can normally be split with a knife at one end and then prised open – you can be quite forceful. You should find cocoons arranged in a line, one after the other. Look for any signs of damage. Rather than a cocoon, some spaces may be empty or filled with tiny mites or their small, pinkish nymphs, or you might see the yellow spaghetti-like excreta of fly larvae. Healthy-looking cocoons can be removed easily from paper or card tubes, but in wooden tunnels they may be stuck to the walls and will have to be eased out with a thin blade. Leafcutter cocoons should be left in their leaf wrapping and lightly brushed clean, whereas mason bee cocoons can be gently wiped with a damp cloth or even soaked in warm water for about 10 minutes and then rinsed clean. Leave the damp cocoons to dry and then put them on some kitchen towel in a sealed mouse-proof plastic container, stored somewhere cool and dark like a garage or shaded outbuilding, or even in a fridge once proper permission has been sought. If they get warm, they may emerge too early.

In spring, put the cocoons in a dark, weatherproof container with a small hole that lets in light. Place it near the original nesting box. When the bees emerge, they will fly to the light and find their way out, hopefully laying new eggs in your nesting box within a week or two. Mason bees will emerge about a month before leafcutters. Some of the better commercially available bee nesting boxes have a built-in drawer for cocoon storage and bee emergence.

If you have removed cocoons from a nesting box, clean it over winter. Scrub wooden nesting grooves with warm soapy water and leave to dry. Replace tubes with new ones and have nesting boxes back in place ready for when bees begin to emerge.

Nesting blocks, or bundles of thin stems from which cocoons cannot be removed, need a different tactic. They can be left in place or can be brought into a cold garage or outhouse for safe overwintering. Place them back outside in early spring. To prevent a build-up of potentially harmful pathogens, nesting blocks or stems should be replaced every two years. The difficulty is in allowing young bees to emerge while preventing the same tubes

from being immediately re-used. As suggested earlier, in spring place nesting blocks or bundles of reed outside in a covered container, so that emerging bees can escape. Make sure that there are replacement nesting materials nearby for them to recolonise.

Nest sites for ground-nesting solitary bees

Most species of solitary bee won't use nest boxes of any kind, preferring to dig or adapt their own tunnels in flat ground, banks, cliff faces or old walls. Creating or providing these conditions is more challenging than making nesting boxes, but the effort is more than worthwhile as they can attract a wide range of species that otherwise rarely nest in gardens. As with bee nesting boxes, success will depend partly on what bee species live locally.

Lawns

If you have a lawn, ground-nesting bees might already be using it. Some species nest in aggregations, in which case you will notice dozens or sometimes even hundreds of bees coming and going, usually from mid-spring to mid-summer (although the late-summer nesting aggregations and mating activity of the ivy bee, *Colletes hederae*, can be quite spectacular). Otherwise, look out for single holes surrounded by freshly dug soil. Watch for a few minutes and you should be rewarded with a fleeting glimpse of a bee zooming out or perhaps arriving laden with pollen, or possibly further excavating her burrow. Tawny mining bees (*Andrena fulva*) and ashy mining bees (*A. cineraria*) commonly nest in lawns, their large size and showy coloration making them quite conspicuous. Many smaller, less easily distinguished ground-nesting species might be seen too, such as the various furrow bees (*Lasioglossum* & *Halictus* spp.). If you find nests, place markers next to each hole so that you can avoid treading on them or mowing over them while the nesting bees are active. Ideally, mow at dusk when flying has finished for the day. Even better, avoid mowing altogether during the nesting season, as discussed in the section on wildflower lawns (see page 243).

ABOVE Female ashy mining bee nesting in sandy ground.

ABOVE Female tawny mining bee burrowing in a lawn.

ABOVE A compacted and overgrown gravel path – ideal conditions for some mining bees, including the tiny green furrow bee (*Lasioglossum morio*), seen sitting on the hosepipe (below).

ABOVE Female sandpit mining bee (*Andrena barbilabris*).
BELOW Sandpit mining bees mating.

Some bees prefer to nest in compacted, dry and sandy soil, or dry loam. Some prefer chalk and some prefer clay. Usually they will only nest in bare or thinly vegetated areas in a sunny location. If you are not fussy about having a perfect lawn and you are sure there are not yet any nests, try mowing a few patches with the mower blade at its lowest setting to scalp the grass (especially if the ground is bumpy). You can also make bald patches by scuffing the surface with the blade of a shovel. Another method is to always walk on the same part of the lawn, compacting the soil and wearing the grass thin.

It's difficult to adapt the soil conditions of existing lawns to make them attractive nesting places, but it's easy to change the soil in borders. Simply clear an area in a sunny spot – typically at the front of a border – and dig in several bags of builders' sand. Keep the area weed-free and in spring keep a look-out for the tell-tale volcano-like mounds that indicate bees are nesting.

Bee banks

You can attract a wide variety of ground-nesting bees to nest by creating mounds of various substrates. Known as bee banks, these resemble rockeries, though they need to be kept relatively free of plants. The key to making them attractive to a range of ground-nesting bee species, as well as other wildlife, is to use as many different substrates as possible to create diverse habitats within a small space. Builders' sand (which is excellent for tunnel building), soft sand, hoggin (a mix of gravel and clay used for making paths), and finely crushed materials including brick, concrete and even ceramics can all be

TIP: Construction of bee banks can be as straightforward or complicated as you like – from simply dumping mounds of substrate in a sunny, unused corner to carefully sculpting something more permanent that fits your overall garden design, perhaps incorporating pathways, borders or, best of all, a pond. Choose a location in your garden that will allow the maximum area to face south or south-east to catch the sun. A crescent shape is ideal as it helps to trap warmth and light. It can be any length, width or height that you like, but the bigger the better.

ABOVE Bee banks made from a range of substrates.

used. Find a local supplier of substrates and see what is available, using as much recycled material as possible. It's surprisingly cheap, although delivery costs can be high.

When the bank is complete, plant flowering annuals and perennials around the edges to provide a source of pollen and nectar. This will attract bees and encourage nearby nesting. Keep the bank itself relatively plant-free so the substrates remain exposed to the sun and available for tunnel-digging. A few scattered plants will help to prevent erosion, but light weeding might be needed to prevent the bank becoming overgrown. Recharge the bank every few years with additional layers of substrate.

WHAT YOU WILL NEED

- Substrate – a range of materials (listed earlier)
- Logs and branches
- A spade and possibly a pick axe or mattock

1. Remove turf and topsoil to a depth of about 15cm in an area the shape and size of your planned bank, plus an additional radius of at least 50cm all around. Pile the turfs to one side for later use.

2. Dig out the cleared space to a depth of about 30cm, placing the spoil in another pile.

3. Stack the removed turfs upside down in the middle of the cleared area to form a central wall or core.

4. Pile the removed soil over the core to increase its height and width. Firm it down well with the back of a spade. You can add any other materials you have to hand, such as old bricks or paving slabs.

5. Pile your substrate materials over the core to form an outer layer that extends to the edge of the dug-out area. The substrate needs to be at least 30cm deep, to give stability and enough depth for bees to tunnel into.

6. Create a range of textures by having some areas covered with a single substrate and others with a mixture of substrate sizes and materials. Mound the substrate quite roughly to make ridges and depressions. This will make micro-habitats with varying levels of moisture, light, shelter and plant cover.

7. Create areas to attract cliff-nesting species by digging away some of the sides to make vertical faces.

8. Logs and branches drilled with holes can be incorporated into the structure to provide stability and additional nesting habitat.

ABOVE The dark-edged beefly (*Bombylius major*).

ABOVE *Colletes daviesanaus* entering nest tunnel in a wall.

Bee banks are good places for spotting one of the more spectacular bee parasites, the dark-edged beefly (*Bombylius major*). These fluffy flies, resembling small, brown bumblebees, zip around the garden and hover over flowers to insert their enormously long proboscis and extract nectar, much like hummingbirds do. You may see them lying on the ground like nesting birds but they will be the filling a pouch with fine grains of soil or sand onto which they will attach their eggs to make them heavier. They are often seen hovering over bare ground as if minding their own business, but are in fact firing their eggs into and around the tunnels of ground-nesting solitary bees. Their developing larvae then eat the host larva and its store of pollen. The adults sometimes even lay their eggs on flowers and their larvae attach themselves to visiting bees and hitch a lift to their nests. Delightful and charming they may be to watch, but they are also just as ruthless as so many other species in the natural world.

Walls and bricks

Soft mortar in stone and brick walls is a favoured nesting location for some bee species. If you have existing old walls, resist the temptation to repoint them. If the walls are not structural you can even scrape out a few areas of sound mortar and replace it with soft, sand-rich mortar, drilling a few holes in it to encourage nesting.

ABOVE Hairy-footed flower bee nest tunnels in the loose mortar of an old brick wall.

If you build new walls in your garden, drystone walls are best for wildlife of all kinds, providing places for various creatures to live, nest and overwinter. Soft mortar between some stones will provide nesting places for solitary bees.

Cob walls made from an unfired mix of clay and soil (known as adobe in the USA) can be highly favoured as nesting sites for mining bees, sometimes being used by large colonies. You can easily make your own cob bricks where bees such as leafcutters (*Megachile* spp.)

and, in particular, hairy-footed flower bees (*Anthophora plumipes*) might nest.

When dry, the cob bricks can be stacked in a sunny position in the garden, ideally at a height of about 1–2 metres – perhaps along the top of an existing wall. Importantly, they need to be sheltered from rain, so cover them with a plank of wood or roof tiles, or build a dedicated shelter. The bricks can also be mixed with other nesting materials in bee nesting boxes of the type described earlier in this section, or can be made directly within their own wooden box which can then be hung on a sunny wall. They should be replaced every few years using the methods already suggested for cavity nesting materials. If the bricks have been used by hairy-footed flower bees and you store them inside over winter, make sure that they are back outdoors by the end of February, when the young bees can start to emerge.

WHAT YOU WILL NEED

- Clay-rich soil
- Builders' sand
- Chopped hay or straw
- Water
- A bucket
- A mould or former
- Sticks or dowels 6–10mm in diameter

1. The main ingredient is clay-rich soil. You might be able to dig this up from your garden, otherwise a perfect material is cricket pitch loam. This is sold in sacks as a dry soil and clay mix and is perfect for making cob bricks.

2. Depending on the size and quantity of bricks you wish to make, mix about three-quarters cricket pitch loam with one-quarter builders' sand.

3. Add several handfuls of chopped straw, which will bind the materials and give the brick more strength.

4. Add water and mix by hand to produce a moist but firm clay mixture.

5. The bricks can be any size or shape you like, but about twice the size of a large house brick is ideal. Make a simple wooden frame to help mould the shape you want, or find any container of the right size to use as a mould. Fill it with the mixture and then turn out the formed bricks.

6. While the mix is still wet, make a series of nesting holes. An ordinary pencil is a good tool for making tunnels of up to about 10mm in diameter. Push the pencil into the soft clay to a depth of 8–10cm and wiggle it about to create a slightly wider hole.

7. Let the bricks dry slowly so that they do not crack. Additional holes can be drilled into the dry cob if needed.

ABOVE Mixing the ingredients for cobb bricks.

ABOVE Making holes in a newly-formed brick.

PART THREE

Gardens for bees

The ever-deepening environmental and climatic crises faced by our planet can sometimes make you despair and question whether your own small efforts, be they recycling, using the car less often, or turning down the heating, can really make any difference. But one way you can have a genuinely positive impact, and even see the results for yourself, is to garden for wildlife. However small and insignificant your garden may seem, you really can make things better, and especially for bees.

A world outside your back door

There are some 23 million gardens in the UK, amounting to about one million acres – much more land than is occupied by our national nature reserves. If even half of these could be made into havens for wildlife, imagine what a difference that would make. Somewhat counter-intuitively, small back gardens can offer richer and more diverse habitat than much of our countryside. No naturalist likes the idea of new houses being built on farmland, but if the average intensively managed arable field was to be replaced with sensitively built homes and gardens, the chances are that local biodiversity would benefit. Interestingly, the trend for more compact gardens seems only to improve matters, often with even more plants being crammed into smaller spaces. Doubling the size of a garden may increase plant diversity by only 25 per cent, according to one study, so smaller gardens can be relatively rich pickings if you are a resident bee. Research has also shown that gardens can have a positive impact on surrounding countryside, acting as pollinator reserves and actually enhancing the productivity of nearby farmland. It is sad that this should be the case, but UK farmland is counted as among the most nature-depleted environments on the planet. Nearby gardens can help to buffer that damage.

The number of species that might make use of our gardens is extraordinary, as revealed in a famous study by ecologist Dr Jennifer Owen. Between 1972–2001, Owen meticulously recorded the wildlife in her remarkably modest family garden in suburban Leicester. Her final tally was 2,673 species. Around 2,200 of these were insects, including 59 species of bees. Sadly, the number and frequency of recordings dropped considerably over that 30-year period, reflecting a worsening environment.

LEFT A mid-summer garden paradise for bees.

PREVIOUS PAGES Honey bee worker with a large pollen load approaches a springtime crocus.

Even so, it is still not uncommon for even urban gardens with a good mix of flowering plants and nesting opportunities to host around 20 species of bees, with some managing as many as 50.

The key to biodiversity is variety of habitat, which is exactly what gardens can offer. Consider the average suburban street with its patchwork of adjoining back gardens; some will be neat and tidy with orderly, flowering borders and lots of exotic plants; some will be more naturalistic, perhaps with more native flowers; others will be unkempt with a variety of uninvited wild plants and possibly some piles of timber or rubble; some will be mostly lawn, and yet others will grow fruit and vegetables. There will be shrubs, ivy-covered walls, and a scattering of mature trees and outbuildings in various states of repair. Although not natural, these many contrasting environments concentrated within a small space can cater for a diverse variety of wildlife, including bees. In a suburban area there might be lots of patches like this within a short flight of one another, perhaps with interconnecting parks, tree-lined roads, railway embankments and allotments, creating a network of valuable habitats.

Many species of bees live and breed within relatively small areas, so your garden could constitute their entire world – for some perhaps the only place they ever use to feed and nest. With that in mind, aim to provide as many suitable flowering plants and nesting habitats as possible. Don't get caught up into thinking that a wildlife garden has to be completely wild to be of any use. There is a sliding scale of wildness, and where you want your patch to sit on that scale will depend on your preferences, time and resources. A garden packed with colourful exotic flowers can be just as wildlife-friendly as one overrun with nettles and brambles. The ideal would be a bit of both. Gardens are highly personal spaces that should allow you to express your creativity and above all make you happy. If you like formal gardens with closely clipped hedges and a striped lawn, that's fine – but take every opportunity to use plants that produce flowers useful to bees. If you want a complete wilderness, that's fine too – although wildernesses do need to be managed to avoid a small number of thuggish plants taking over completely.

Whatever you think you want, be in no hurry. Observe the bees that already visit your garden and plant more of what they seem to like. Look for other species of bees in local parks and gardens, and if you don't already have the right plants for them, put some in – they will probably find and use them. Your local Wildlife Trust is likely to have an entomologist who can tell you which bees are found locally and whether any are in particular need of support, and you can plant accordingly. A garden is not just a place for bees to feed but also to reproduce, so provide as many different nesting opportunities as you can, using the techniques already described in Part Two.

Gentle gardening

If bees and other wildlife are your priority, then sustainability should be at the forefront of everything you do in the garden. General measures include: avoiding the use of peat-based composts, reducing and recycling plastic waste such as plant pots, conserving water, sourcing plants grown in local nurseries or through local sales and seed swaps, making compost and leaf mould, and using no-dig techniques wherever you can. Your garden needn't be messy, but allow a little untidiness here and there; the odd pile of dead leaves or stack of gently decaying logs will make quite a difference to wildlife and can be attractive too.

Above all, garden organically. No pesticide can be considered safe for bees. Insecticides, obviously, are highly dangerous, killing insects indiscriminately, whether blackfly or bumblebees. But increasingly, evidence shows that fungicides and herbicides, and in particular some of the ingredients that they are mixed with, can be highly damaging to all wildlife and to the environment in ways we are only just beginning to understand. A healthy, well-balanced garden will thrive without chemicals. Using them only puts the balance off-kilter, fracturing the natural harmony you have striven to achieve in a way that is difficult to repair.

Gardening organically includes how you feed your plants. Good soil and plenty of compost are almost all that most plants need, but a little boost to long-flowering bee plants, those in pots or anything particularly hungry, can keep them going a little longer. There's no need to buy manufactured oil-based fertilisers when something better can often be made at home. One of the best all-round feeds is smelly but nutritious comfrey fertiliser, easily made from a beautiful plant which should be grown in every bee garden (see page 280).

The Victorians used lead arsenate freely in their gardens as an insecticide. That now seems like absolute madness, but what will future generations think when they look back at what we have been doing in our own gardens – particularly now that we know better?

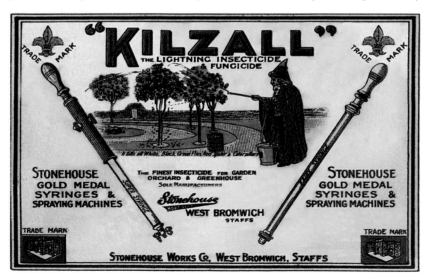

LEFT The dark art of pesticide use has been practiced in gardens for over a century, as this advert from the 1920s shows. The benefits of organic gardening are becoming increasingly obvious.

Seeing the difference

I wrote earlier that gardening for bees can make a difference that can be seen. Almost as soon as you plant flowers of the right kind, bees of some sort will arrive and make use of them. Plant more and you will get more bees, and different species too. Provide nesting sites and the number of bees will increase year on year, with populations often growing noticeably and requiring yet more flowers and more nesting sites. After several years you might begin to get the sense that the bee population of your garden is outgrowing the habitat you can provide for them – which is when you might want to encourage friends and neighbours to join your enterprise. What's more, lots of other wildlife comes for free; by creating a rich and varied habitat designed for bees you will by default attract and support a wide range of other creatures, including wasps, beetles, hoverflies, moths and butterflies, not to mention birds, amphibians and small mammals. All of this life can be buzzing, fluttering, thrumming and tweeting just outside your back door, and all it takes is a few well-chosen plants and some sympathetic gardening techniques.

Plants and bees

Plants provide bees with almost all the resources they need to sustain them – most importantly the pollen and nectar produced within flowers. Pollen contains the male genetic material of a plant and must be transferred to the female parts of a flower to achieve fertilisation and the production of seeds. The process of transferring pollen is called pollination and can be achieved in different ways. Some flowers rely on air currents to waft pollen to where it is needed – a process called anemophily. Others rely on insects to transport it – a process called entomophily. Some hedge their bets and use a bit of both, which is called ambophily.

The structure of flowers differs depending on how they receive pollen. Wind-pollinated flowers are simple, designed for the efficient collection of airborne pollen as it drifts past. Those that rely on pollinators have evolved various elaborate lures to attract bees, butterflies, beetles, moths and other creatures. These include colour, scent, shape and, crucially, nectar. As pollinators like bees try to reach this sugary solution they brush against the pollen, which sticks to their

BELOW This preening solitary bee (*Andrena* spp.) has collected tansy pollen on its feather-duster-like leg hairs.

hairy bodies. Like miniature feather dusters they then transfer this pollen to the next flower that they visit, thus achieving pollination.

Bees collect nectar because it is primarily a source of carbohydrates. Some is consumed immediately to give them energy for warmth or flight, and some is taken back to the nest to feed their young. Only honey bees can convert nectar into honey for long-term storage.

Pollen is a source of protein as well as various fats and minerals. Bees pick up and transfer pollen from flower to flower almost accidentally when seeking nectar, but females with young to feed will actively gather it in large quantities. In an amazing example of the harmonious relationship between bees and flowers, the hair on a flying bee builds up a positive electric charge which, when the bee lands on a flower, attracts the negatively charged pollen like iron filings to a magnet. Bumblebees and honey bees brush the pollen off their hairy bodies, stick it together with a little nectar and pack it into pollen baskets on their hind legs for the journey home. Some solitary bees collect pollen on a hairy brush on the underside of their body, called a scopa, while a few carry pollen to their nest by swallowing and later regurgitating it.

Different kinds of pollination

Some plants are able to self-pollinate, using their own pollen to fertilise themselves without any external input. However, other plants need to cross-pollinate, meaning that pollen from another plant of the same species must be used. In this case there are various mechanisms to

ABOVE Some plants have separate male and female flowers. The simple female flowers on hazel are designed to catch the airborne pollen produced by the male catkins.

ensure that pollen from the same flower or plant cannot cause self-fertilisation. This can involve male and female parts of a flower maturing at different times, or physical barriers preventing pollen from fertilising the flower that it came from. The majority of flowers contain both male and female reproductive organs, but some plants are monoecious, meaning that they grow separate male and female flowers. Others are dioecious – that is, male and female flowers are produced on separate plants.

Flower structure

Flowers come in an astonishing variety of shapes and sizes, but apple blossom is a good example of a flower in which all of the reproductive elements can be easily seen and understood (see image opposite). Around the outer edge of the flower are the sepals – leaf-like structures that covered the growing bud before peeling back to reveal the flower. There are five pink or white petals that form a bowl, known as the corolla. The petals are like flags, advertising the presence of the flower to pollinators – a subtle scent helps in this task. In the centre of the flower there is a ring of stamens, each a fine filament with a bud-like anther on top. The anthers open (dehisce) to release pollen grains that have formed inside them. There is an inner cluster of five styles, each tipped with a stigma, the female receptive organ. A grain of pollen reaching a stigma will sprout a long, hair-thin tendril that grows down inside the style until it reaches the ovaries in the base of the flower, delivering the male sex cell to complete fertilisation. If all of the ovaries in an apple flower are fertilised, they will swell, eventually forming seeds surrounded by the flesh of the apple. Inadequately fertilised flowers will result in apples in which some of the seeds don't develop, producing fruit that are lopsided and misshapen – though still just as tasty.

Nectar

An important part of flower anatomy not mentioned above are the nectaries. These are tiny glandular organs usually at the base of the styles and stamens, which secrete nectar. The position, size and productivity of nectaries vary enormously from species to species and under different growing conditions. It is while seeking nectar that bees and other pollinators pick up and transfer pollen from flower to flower. Flowers that rely

LEFT Apple blossom showing a typical flower stucture consisting of petals, stamens and styles tipped by stigmas.

BELOW A perfectly formed apple, the result of complete pollination.

on wind pollination do not require insects to help them and so produce no nectar, only pollen. However, they are often visited by bees anyway, as their pollen is usually plentiful and easy to collect.

Although most nectar is produced within flowers, some plants have other outlets. These 'extrafloral' nectaries are usually small, dark spots on stems or under leaves. It is not fully understood why plants should secrete nectar from anywhere other than their flowers, but bees certainly make good use of extrafloral nectaries on plants including cherry laurel (*Prunus lauroceracus*), vetches (*Vicia* spp.) and cherries (*Prunus* spp.). Honey bees can produce a good crop of honey from the nectar

secreted by the extrafloral nectaries on broad and field beans (*Vicia faba*) when grown on an agricultural scale.

Plants can also supply sugar to bees indirectly. Aphids and scale insects living on plants pierce them with their sharp mouthparts and suck out sap. The sap contains tiny amounts of nutrients and quite a lot of sugar. To obtain sufficient nutrients, these tiny sucking creatures process a lot of sap and are left with an excess of sugar. This is secreted onto leaves – and possibly onto your car if you park beneath a tree full of aphids. Called honeydew, this secretion is licked up by bees, often in the morning when the moist air liquifies it. Later in the day, it can be too sticky to remove.

ABOVE Aphids and scale insects leave sugary deposits on leaves.

ABOVE This cutaway oilseed rape flower shows a droplet of nectar secreted from nectaries at the base of its styles and stamens.

ABOVE A honey bee working the extrafloral nectaries on a broad bean plant.

How flowers attract bees

Plants have evolved a range of attractive qualities in their flowers to lure bees and other pollinators to them. Many are brightly coloured, usually with colours in parts of the spectrum that can best be seen by bees. Bees' eyes are different from those of humans, being more sensitive to short-wave light and less sensitive to long-wave light. This means that they can see ultraviolet light, but to them a red flower appears black. Many wild red flowers have evolved to be pollinated by other creatures, such as the wild red salvias and fuchsias that are pollinated by hummingbirds. Predominantly bee-pollinated flowers tend to be in shades of white, yellow or blue. However, many flowers, including some red ones, have areas that reflect ultraviolet light which we cannot see but which show bees exactly where to land and look for nectar. These are called nectar guides. Some flowers also go through colour phases, changing colour to indicate when they are ready for pollination or when they have already been pollinated. Forget-me-nots (*Myosotis sylvatica*) are a good example of this. At the centre of the powder-blue ring of petals is a fleshy yellow corolla which invites bees to visit the flower and collect nectar, while at the same time picking up or depositing pollen. Once the flower has been pollinated the ring turns from yellow to creamy-white, indicating that bees should spend their time visiting one of the other many forget-me-not flowers likely to be nearby.

The flower's shape and size help to indicate what kind of bee it is best engineered to serve. Bees vary greatly in size, from the smallest solitary bees at just a few millimetres in length, to the biggest bumblebees, which are many times larger. Importantly, their tongues vary in length, allowing them to access nectar that is offered at different depths within flowers. Small solitary bees have tongues just a few millimetres long, whereas some bumblebees can have a tongue as long as 15mm. Honey bees have quite a short tongue, at about 7mm, while the hairy-footed flower bee (*Anthophora plumipes*) has a tongue as long as 20mm. Flat-topped flowers with many tiny florets, such as tansy and yarrow, are favoured by bees with short tongues. These species can walk across the top of the structure, dipping into each of the shallow florets to reach the droplets of nectar they contain. Many of the open, cup-like flowers are suitable for short- or medium-tongued bees, such as honey bees and most of the bumblebees. Tubular flowers that have their nectar at the bottom of a deep, narrow, trumpet-shaped corolla, such as foxgloves and penstemons, are generally worked by bees with the longest tongues. In the UK, the most common longer-tongued bumblebees are the common carder bee (*Bombus pascuorum*), which has a medium-length tongue and the garden bumblebee (*B. hortorum*), which has a long tongue. However, some short-tongued bees such as the buff-tailed bumblebee (*B. terrestris*) have developed a cunning way to get to the nectar of tubular flowers – they bite a hole in the base of the corolla and poke their tongue through it. Once made, these holes will be used by other species of bees as well, robbing the nectar from a flower without pollinating it.

Some flowers are designed so that only the biggest or strongest bees can push their way in to get at the nectar. This is normally linked to a mechanism that forces the bee to pollinate the flower. For example, salvia flowers have a protruding bottom lip which offers a landing platform for bees. As they enter, the lip swings upwards and pushes the bee onto the anthers or stigma, to accurately deposit or remove pollen from a precise spot on the bee's body.

ABOVE When seen by a bee, the ultraviolet markings on a dandelion flower indicate where to find nectar.

ABOVE A buff-tailed bumblebee robbing nectar through holes nibbled in the base of an escallonia flower.

ABOVE The corolla of a forget-me-not flower changes from yellow to white after pollination.

ABOVE The anthers of this salvia dab pollen onto a bee when it mechanically activates a trigger.

ABOVE Flowers with many shallow florets attract bees with the shortest tongues, like this tiny *Sphecodes* solitary bee on tansy.

ABOVE Very deep flowers like this lobelia can only be accessed by bees with very long tongues such as the garden bumbleee (*Bombus hortorum*).

What to plant for bees

The following pages contain information and lists of plants recommended to attract and help sustain bees. The suggested plants all have flowers that in one way or another conform to the following criteria that need to be considered when planting a bee garden.

Shape, size and colour

Based on the way they present themselves, different kinds of flowers attract different kinds of bees. Try to have a mixture of large and small flowers, tubular and bowl-shaped flowers, and composite flowers – those with hundreds of tiny florets at their centre, such as heleniums and asters.

Many garden plants have been selectively bred to produce flowers that are bigger, brighter or more elaborate than their wild forms. Sometimes this can mean a better offering for bees, but frequently what makes them attractive to bees is lost. Some no longer have their appealing scent or colour, some are sterile so have no need to produce pollen and nectar, and some are so radically different in shape that bees can no longer reach whatever nectar or pollen they may contain. A good example is the dahlia, which is naturally an open, daisy-like flower much loved by bees, but which has been bred into flower forms including double, pom-pom, ball, anemone, waterlily and cactus-shaped – many of which are of no use to bees. Where possible, stick to old-fashioned, un-hybridised, single-flowered plants – the closer to their wild form the better.

Origin

Recent research shows that it doesn't matter where a plant comes from as long as its flowers provide accessible pollen and nectar – and bees certainly don't discriminate. Many non-native plants in fact provide resources at times when few native plants are in flower, particularly in late summer. Some are better suited to warmer or drier climates and will produce nectar in droughts when other plants have given up; dry-garden plants are likely to become increasingly important as gardens and bees adjust to a changing climate. Native

LEFT Forest cuckoo bumblebee (*Bombus sylvestris*) on drumstick alliums. Alliums of all kinds are tremendously attractive to bees.

BELOW Flowers like this pom-pom dahlia are of no use to bees.

RIGHT Bees don't care whether a plant is native, as long as they can access pollen or nectar. This phormium is native to New Zealand.

ABOVE Drifts of *heleniums*, *echinops* and *echinacea* provide a banquet for bees.

plants do have an important role to play, though, in cases where they have a close relationship with a particular species of bee. For example, white bryony (*Bryonia dioica*) is used by the bryony mining bee (*Andrena florea*). To keep pollinators happy in your garden, plant a range of native and non-native flowers so that there is always plenty to choose from.

Quantity

Bees use a lot of energy when flying, beating their wings around 300 times per second. Reducing the distance between one food source and another can make a great difference to their efficiency and prosperity. Where possible, plant flowers of the same type in large clumps or drifts, with relatively small gaps between plants that flower at the same time. In fact, planting in drifts attracts relatively more bees to your garden. For example, one or two lavender bushes will attract a handful of bees, but twice as many bushes can attract more than twice as many bees. Some plants simply won't attract any bees unless they are planted in quantity. Forget-me-nots, for example, go unworked where there are only a few flowers, but are very popular when planted en masse. Try not to waste any opportunity to provide forage for bees; every tree, hedge,

border or planted pot in your garden has the potential to be a source of nectar and pollen, so in each case choose plants that produce flowers useful to bees. If you notice a particular species or cultivar that seems to attract bees more than others, plant more of the same, perhaps by collecting seed, dividing or taking cuttings.

Flowering time

It's important that nectar and pollen are available when bees need them, and several periods are particularly crucial. Late winter and early spring are dangerous times for early-emerging queen bumblebees, which need nectar to give them the energy to fly and keep warm. Nectar-rich mahonia or spring-flowering bulbs can be life savers. Honey bees also need pollen and nectar from early spring onwards. June can be a difficult time, as many spring flowers have ended but summer flowers have not yet bloomed. Beekeepers call this period the June gap and it can lead to the death of colonies, so encourage anything that flowers in your garden at this time. Summer is a long season when there can be many bees in search of food. Try to have plants that will flower for a long period or that are replaced by something else when they go over, ensuring a continuous supply. Many plants will flower all summer, and even until the first frosts, if they are regularly deadheaded and fed as necessary. Another method is to give a 'Chelsea chop' to a portion of your plants, cutting back herbaceous perennials in late spring to induce bushier growth and later flowering. You can also cut back early-flowering perennials after their first flush of flowers, usually in about mid-June, to induce a second flush later in the summer. Autumn is the last chance for insects that overwinter to fill up on nectar, so anything that flowers late in the year should be encouraged. Crucial at this time is flowering ivy, perhaps the most important pollinator plant of all.

Recommended plants

The lists on the following pages are based on recommendations made by expert beekeepers, entomologists and personal observations over many years. I have avoided recommending plants just because they frequently appear on lists of plants for bees; if I have not regularly and consistently seen them being used, or know them to have a special use for bees, they have not been included. In fact, it can be difficult to determine which plants are best used by bees, since several variables affect the production of nectar, including time of day, soil quality, temperature, rainfall and so on. Some plants that are supposedly good for bees can flower year after year without bees apparently showing any interest before suddenly being used enthusiastically – perhaps because conditions are just right or because something more attractive nearby has recently been removed. Equally, plants not thought to be of use to bees will sometimes be found inexplicably attractive. No list of plants for bees can be totally comprehensive, and those in this book are no exception. Perhaps the best advice I can give is to visit gardens, particularly those nearby, and observe what flowers bees are using, and if you like them, plant them. There are several excellent apps which, when used with a smartphone camera, are very helpful for identifying garden plants.

Apart from a few very significant plants, I have tried to avoid duplication, so most plants are included on one list only. It hasn't always been easy to decide the best place – should Sichuan pepper be listed as a fruit, herb or shrub, for example? Since some plants have several common names, in most cases I give the scientific name. Note that the qualifier spp. denotes that there are several

ABOVE Knapweed is a top-rated three-star recommendation, here being visited by a spined mason bee (*Osmia spinulosa*).

species within a genus. For example, *Teucrium* spp. means that there is more than one species of *Teucrium* that is good for bees. In most cases, I have listed family and species names, referring to specific cultivars only where they are known to be particularly attractive to pollinators or have other benefits. In all cases, single-flowered varieties are best.

As this is a book about bees more than gardening, I have included little information about plant cultivation other than where it relates to methods that might make a specific plant or planting situation more useful and productive for bees. Refer to good gardening books or your nursery for full information about plants and their requirements.

I have used a star system, awarding up to three stars according to the usefulness of plants to bees, as follows:

* A useful bee plant, helpful for a small range of bees, for a limited period or in specific circumstances.

** An excellent bee plant, appealing to a range of species in most circumstances.

*** An essential bee plant, which reliably produces pollen and nectar in large quantities or is used by a wide range of species. Plant this if you can.

The winter garden

Deep mid-winter is no time for bees to be active. Nonetheless, on milder days honey bees often do fly – emptying their bowels on so-called cleansing flights, collecting water to dilute honey stores and gathering nectar and sometimes pollen if available. Bumblebees too can sometimes be seen in winter. Typically, they hibernate in north-facing sites, a tactic to avoid being tricked by several sunny days into thinking that it's spring. However, as climate change brings milder winters, bumblebee queens are increasingly seen flying at times unthinkable several decades ago. Buff-tailed bumblebees in particular are now regularly seen in winter, with some queens even emerging from a very short sleep to start new nests in late autumn or early winter. Their small workers can be seen seeking sustenance in winter gardens, sometimes even when there is snow on the ground. In early spring it is not uncommon to find struggling queens that have been caught out in the cold or rain. If you find one, bring her inside, warm her up and offer her a thick syrup made of white table sugar. These queens often revive

quickly and fly away. Having saved a queen in this way, you may have ensured the survival of a whole generation of bumblebees.

As winter slips into early spring, honey bee queens gradually increase their egg-laying and the colony begins its incremental build-up to summer. Workers must feed the young, and will use stores from within the nest if they can. However, it is important that they can find nectar and in particular fresh pollen at this time, flying from the nest whenever the weather allows to hunt for crucial supplies. This is a dangerous time; early spring can have some of the fiercest weather, with the potential to confine honey bees and bumblebees to their nest for weeks, with young that need to be fed and kept warm. It is the time when most colonies fail.

Winter-flowering plants need to be pollinated at a time when few insects are available. To find customers, many advertise themselves especially well by releasing strong perfumes that travel far, alerting the few available pollinators to their presence. Winter blooms are therefore among the most sweetly scented garden plants. Another ingenious tactic is used by hellebores – mainstays of the winter garden. The drooping bell

ABOVE In mid-January a honey bee forages on *Clematis cirrhosa* 'Lansdowne Gem'.

flowers of hellebores act as umbrellas, keeping their long-lasting supply of pollen dry. They also produce copious amounts of nectar which is stored in deep vases produced by their folded petals (the outer petals of the flower are in fact sepals). The clever part is that the flowers are centrally heated; yeasts living in their nectar ferment and produce heat, warming the inside of the cloche-like flowers by as much as 3°C. *Helleborus niger*** is the snow-white Christmas rose that, with warmer winters, might even see some custom on Christmas Day. Most commonly available garden hellebores are hybrids of *H. orientalis*** and will flower in a range of delicate shades from about mid-winter, but the native *H. foetidus***, with demure green flowers subtly rimmed with crimson, will bloom later, extending the season. They are all highly attractive to bees and a must for the winter garden.

An unusual but striking shrub, *Stachyurus praecox***, bears long racemes of pale yellow bell-shaped flowers that are loved by honey bees and early bumblebees. *Praecox* in a plant's binomial name means 'early', and the common name of this shrub is early stachyurus – it sometimes flowers as early as January.

ABOVE Struggling bumblebees can be revived with sugar syrup.

Late winter and early spring flowers are among the most important resources that gardeners can provide for bees; they really can mean the difference between life and death. They also bring delightful touches of life, colour and scent at a time when there can be little else to tempt gardeners, or bees, out into the cold.

ABOVE *Helleborus foetidus*.

ABOVE *Stachyurus praecox* providing early nectar for a honey bee.

Recommended winter- and early spring-flowering plants for bees:

- Crocus (*Crocus* spp., especially *C. tommasinianus*)***
- Cornelian cherry (*Cornus mas*)**
- Coronilla (*Coronilla valentina*)**
- Daphne's (*Daphne bholua, D. mezereum*)***
- Edgeworthia (*Edgeworthia chrysantha*)*
- Hazel (*Corylus avellana*)**
- Hellebores (*Helleborus* spp.)***
- Mahonia (*Mahonia* spp. Recommendations: *M. × media* 'Winter Sun' and *M.* 'Soft Caress' for small gardens/pots)***
- Viburnums (*Viburnum farreri, V. tinus*)**
- Snowdrops (*Galanthus* spp.)**
- Sweet box (*Sarcococca* spp.)***
- Winter aconites (*Eranthis hyemalis*)***
- Winter clematis (*Clematis cirrhosa*)**
- Winter-flowering cherry (*Prunus × subhirtella* 'Autumnalis')***
- Winter heather (*Erica carnea*)**
- Winter honeysuckle (*Lonicera fragrantissima*)**
- Wintersweet (*Chimonanthus praecox*)*
- Yellow nonea (*Nonea lutea*)**

LEFT TO RIGHT FROM TOP:

THIS PAGE Yellow nonea; Hellebore; Winter heather, Mahonia.

OPPOSITE PAGE Hazel, Winter-flowering cherry; Sweet box, Winter honeysuckle; Daphne, Winter clematis.

Bulbs for bees

There is a tendency to think of bulbs as flowering only in late winter and early spring – perhaps because by then we are always so pleased to see their cheery colours and also because they herald the start of the new gardening – and beekeeping – year. However, bulbs can produce flowers in every season, and in every season there are bulbs that can provide something of use to bees. Here I also include corms, rhizomes and tubers within the broad category of bulbs.

Having said that bulbs can flower throughout the year, it is in late winter and early spring that they are most useful to bees; at this time bumblebee queens and foraging honey bees are desperate for any supplies of pollen and nectar they can find. The first bulbs to flower are usually snowdrops (*Galanthus* spp.)**. Honey bees visit these mainly for pollen, and it can be a surprise to see them emerging from the pure white flowers with legs loaded with deep orange pollen. There are hundreds of varieties of snowdrop but, as always, the single-flowered types are best for bees. Bulbs of *G. nivalis*, the simple common snowdrop, are cheap in autumn but they are better bought and planted in spring with their leaves and stalks still attached 'in the green'. Once established, they will spread by seed or can be easily divided every few years. The giant snowdrop (*G. elwesii*) has larger flowers and better accommodates early-emerging queen bumblebees. Although snowdrops are a harbinger of spring, they can also flower in autumn; *G. reginae-olgae* produces leafless flowers in about October. Originating from Greece, it does best somewhere sunny.

At about the same time as late-winter snowdrops come the sunshiny buttercup blooms of winter aconites (*Eranthis hyemalis*)***. Plant these somewhere that catches winter sun and on warm days they will be feverishly worked for both pollen and nectar by honey bees and the occasional queen bumblebee. Their waxy leaves and petals often hold droplets of water which, when sun-warmed, honey bees will harvest too. They can be in flower for as long as six weeks. Next come the crocuses, which are especially welcomed by honey bees who by about March are desperate for every grain of pollen they can find. There are many crocus varieties and by planting a selection they can be in flower for more than a month, with the later flowers being visited by early-emerging solitary bees such as the tawny mining bee (*Andrena fulva*). Autumn-flowering crocuses (e.g. *Crocus sativas*, also *Colchicum* spp.) may also be visited for pollen by honey bees. Snowdrops, aconites and spring crocuses are all worth planting around beehives. They will provide nearby resources when these are needed most, and flowering will be over by the time you regularly visit the bees and trample the ground around their hives.

Among the best known of spring-flowering bulbs, daffodils and tulips sadly aren't of much use for bees. They may provide a little pollen, but if you want to plant bulbs for bees, there are better alternatives. The

ABOVE Snowdrops, an important early source of pollen.

ABOVE Crocus and winter aconites

various species of grape hyacinth (*Muscari* spp.)** differ considerably in habit, and not all are as invasive as the notorious *M. neglectum*, which is loved by bees – even when pushing its way up through your driveway. A similar deep blue colour, but non-invasive, are the lovely spring squills, of which Siberian squill (*Scilla siberica*)** is the most commonly planted. The bell-shaped flowers are not dissimilar in appearance to those of bluebells, which are among the last of the spring bulbs to flower, usually in April. Preferring dappled shade, bluebells (*Hyacinthoides non-scripta* – avoid the Spanish variety)** aren't generally considered a good bee plant, but those in sunny spots can be well worked by some species of bumblebees and solitary bees, and occasionally by honey bees which collect the blue-white pollen from within the bells or push their probosces between the overlapping petals to access the nectar.

Spring and summer overlap with the flowering of the first alliums (*Allium* spp.)***. There are many of these ornamental onions, and all are loved by bees. Early to flower is *A. karataviense*, which is short but has a very large flower. This can be followed by the very commonly grown 'Purple Sensation', which is beautiful and dramatic but can become a bit of a weed. For height, plant *A. macleanii* 'His Excellency', which can reach 1.5m, or for blooming enormity *Allium* 'Globemaster' has flower clusters up to 25cm across. To make the bees in your garden hum with contentment, plant the elegant Sicilian honey garlic, *A. siculum*, and watch them circling as if on a nectar carousel. The last to flower is the drumstick allium, *A. sphaerocephalon*, which blooms in July and well into August.

Although now very trendy, dahlias have always been in fashion as far as bees are concerned. What bees don't like, though, are the many hybrid forms that may have neither pollen nor nectar, or whose complex flowers don't allow access to where these are produced. Simple, daisy-shaped dahlias on the other hand are heaven for bees, and it is not uncommon to see several species at once working their way around the nectar-rich florets. Try any of the single-flowered peony, mignon or orchid dahlias. *Dahlia* 'Bishop of Llandaff' and other varieties in the Bishop series are loved by bees.

Although mentioned here because they are most frequently grown from tubers, dahlias can easily be grown as annuals from seed. Sow under cover in February and plant out after the first frosts. The only

ABOVE Grape hyacinth.

ABOVE *Allium hollandicum* 'Purple Sensation'.

ABOVE Sicilian honey garlic.

catch is that, unlike those grown from tubers, flowers grown from seed will vary in colour. Keep the ones you like by lifting and storing the tubers at the end of their first summer. If growing from seed, try the seed mix called 'Bishop's Children'.

Other suggested flowering bulbs for bees:

Winter and spring

- Camassia (*Camassia* spp.)**
- Crown imperial (*Fritillaria imperialis*)**
- Glory of the snow (*Scilla luciliae*)*
- Snake's head fritillary (*Fritillaria meleagris*)**
- Star of Bethlehem (*Ornithogalum* spp.)*

Summer

- Agapanthus (*Agapanthus* spp.)**
- Crocosmia (*Crocosmia* spp.)**
- Angel's fishing rods (*Dierama* spp.)**
- Foxtail lily (*Eremurus* spp.)***
- Peruvian lily (*Alstroemeria* spp.)**
- Society garlic (*Tulbaghia violacea*)***

LEFT TO RIGHT FROM TOP:

THIS PAGE Scilla; Society garlic, Crocosmia; Agapanthus, Snake's head fritillary.

OPPOSITE PAGE Snowdrop; Peruvian lily, Foxtail lily.

Annuals and biennials

Many plants of use to bees are annuals – meaning any plants that germinate, grow, flower and set seed in the same growing season. Annuals are tremendously useful for the gardener, being able to fill beds, pots, tubs and hanging baskets as well as plugging unexpected or inconvenient gaps in the garden. Some annuals can be planted under cover late in the year to ensure early flowering the following season, while successive sowings can mean that flowers are produced constantly throughout the summer.

Annual bedding plants are readily available from supermarkets and garden centres, but with a few exceptions they are rarely good bee plants. Begonia, pelargonium, pansies, petunias and others, even if promoted as good for bees, very often are not. But if you have a greenhouse or an accommodating windowsill, a few low-priced packets of seed can produce hundreds or even thousands of plants. This affordability, combined with the single-season nature of annuals, means that there is great scope for experimentation. For the bee gardener this means trying different species, cultivars and combinations every year to discover what works well for different kinds of bees.

Biennials differ from annuals in that they grow fast from seed developed in one season to produce plants that survive the winter and then flower, set seed and usually die in their second year. The great advantage of biennials is that, after having endured winter, they grow rapidly in spring and can flower earlier than many annuals, making them useful at a time when there are fewer flowers in the garden for bees. A classic example is the common wallflower (*Erysimum cheiri*)***. These used to be sold wrapped in damp newspaper by greengrocers. If put in the ground over winter, they will produce wonderful colour and scent as well as nectar and pollen from mid-spring the following year and then are usually replaced by summer bedding. For flowering value, however, perennial wallflowers can't be beaten. These shrubby evergreen perennials can flower for a full eight months of the year or more, and are a good alternative to lavender. *Erysimum* 'Bowles's Mauve' seems best liked by bees and is widely available.

LEFT *Cirsium rivulare.*
RIGHT Giant echium.

Many annuals and particularly biennials self-seed very well, popping up around the parent plants and flowering the following year. Some, like forget-me-nots (*Myosotis* spp.)**, become self-perpetuating. It is also possible to collect and save the seed of many plants for later sowing.

Strictly speaking, annuals and biennials flower for just one season before setting seed and dying. Here I have included some species that can be grown from seed and will flower in their first or second year, but which might last several years or become perennial. Milder winters, especially in the south, mean that many tender plants last longer than once they did. Mild weather also makes it possible to reliably grow some tender plants that once were more challenging. For bees, these would have to include the spectacular giant viper's bugloss (*Echium pininana*)***, a biennial native to the Canary Islands that can grow a 5m spire of tens of thousands of nectar-rich flowers. Grow some, give them to your friends and have a competition to see who can grow the tallest – it will make a lot of bees very happy. For those with more modest ambitions, the native viper's bugloss (*E. vulgare*)*** is also a star performer.

Recommended annuals for bees:

- Borage (*Borago officinalis*)***
- California poppy (*Eschscholzia californica*)**
- Cerinthe (*Cerinthe major*)***
- Cornflower (*Centaurea cyanus*)***
- Cosmos (*Cosmos bipinnatus*)**
- Dahlias (suggest: 'Bishop's Children' seed mix; see page 227)***
- Gaillardia (*Gaillardia* spp.)**
- Garden heliotrope (*Heliotropium arborescens*)**
- Godetia (*Clarkia grandiflora*)**
- Chinese lantern (*Abutilon × hybridum*)**
- Love-in-a-mist (*Nigella damascena*)**
- Mignonette (*Reseda odorata*)***
- Mexican sunflower (*Tithonia rotundifolia*)**
- Nasturtium (*Tropaeolum* spp.)**
- Phacelia (*Phacelia tanacetifolia*)***
- Poached egg flower (*Limnanthes douglasii*)**
- Pot marigold (*Calendula officinalis*)**
- Purple ragwort (*Senecio elegans*)**
- Queen Anne's thimble (*Gilia capitata*)**
- Snapdragon (*Antirrhinum majus*)**
- Sunflower (*Helianthus annuus*)***
- Spider flower (*Cleome spinosa*)**
- Sweet scabious (*Scabiosa atropurpurea*)***
- Tickseed (*Coreopsis* spp.) **

THIS PAGE Dahlia (top) and Cornflower (bottom).
OPPOSITE PAGE, LEFT TO RIGHT FROM TOP Borage, Cerinthe, Cosmos; Pot marigold, Phacelia, Mexican sunflower; Spider flower, Sunflower, Poached egg flower.

Recommended biennials for bees:

- Anchusa (*Anchusa* spp.)***
- Arabis (*Arabis* spp.)**
- Black-eyed Susan (*Rudbeckia hirta*)***
- Canterbury bells (*Campanula medium*)**
- Common wallflower (*Erysimum cheiri*)***
- Cotton thistle (*Onopordum acanthium*)***
- Evening primrose (*Oenothera biennis*)**
- Foxglove (*Digitalis* spp.)***
- Hollyhock (*Alcea rosea*)**
- Honesty (*Lunaria annua*)***
- Larkspur (*Consolida ajacis*) **
- Plume thistle (*Cirsium rivulare*)***
- Siberian bugloss (*Brunnera macrophylla*)***
- Sweet rocket (*Hesperis matronalis*)***
- Viper's bugloss (*Echium vulgare* 'Blue Bedder')***
- Wild mignonette/weld (*Reseda lutea*)***

THIS PAGE Honey bees on *Erysimum* 'Bowles's Mauve' (top) and with bumblebees on Anchusa (bottom).

OPPOSITE PAGE, LEFT TO RIGHT FROM TOP Sweet rocket, Evening primrose, Foxglove; Hollyhock, Cotton thistle, Forget-me-not; Black-eyed Susan, Viper's bugloss, Honesty.

The perennial garden

Of all the plant groups, herbaceous perennials are the most versatile. They are widely available and generally easy to grow in a range of conditions. Varying greatly in size, shape and colour, they can flower from early spring until the first frosts, and sometimes beyond, supplying many months of forage for bees and pollinators of all types. After flowering, many lend a garden winter drama and structure, as well as seedheads to feed the birds.

There has been a huge revival of interest in the use of herbaceous perennials over the past two decades, thanks in part to the work of the Dutch planting designer, Piet Oudolf. Pioneering the use of long-flowering perennials and grasses in large undulating drifts, Oudolf has created a naturalistic garden aesthetic that has been hugely influential. For bees, an Oudolf-style garden must seem like an enormous buffet, with huge portions of irresistible food laid out at close intervals. Oudolf's ideas are usually achieved in landmark locations, but the effect can be reproduced on a more modest scale at home – the secret is to use fewer varieties of flowers in larger groups or drifts. The approach works particularly well with island beds – inverting the traditional arrangement of a lawn with flowerbeds around the margins. The beauty of perennials is that they can be used in almost any situation, so if expansive drifts are off the cards, you can still achieve beautiful bee-enticing displays on balconies, in pots, and in mixed cottage garden-style borders.

Herbaceous perennials are usually bought as pot-grown specimens, but can be expensive if planted in large groups as suggested above. Small plants, plugs and bare-rooted specimens are much cheaper and they quickly catch up. Once established, some are easily reproduced from seed, while others take well from cuttings or can be divided. Many self-seed very well and a large part of the work in a perennial garden can be editing out what nature has provided for free.

Some favourite herbaceous perennials for bees

The cheerful pink, blue or white flowers of lungwort (*Pulmonaria* spp.)*** usually appear from about early March. In my garden, where it self-seeds in gravel paths, I use it to gauge the progress of spring. The first, occasional visitors are buff-tailed bumblebee queens (*Bombus terrestris*). The funnel-shaped flowers are a bit too deep for their short tongues, but they persist anyway. Around mid-March, the hairy-footed flower bees (*Anthophora plumipes*) appear; first the gingery-coloured males, followed about a fortnight later by the darker females. Then the first early bumblebee queens (*B. pratorum*) begin to arrive, always surprising me with their diminutive size but vivid colours. The final arrivals are the first small buff-tail or early bumblebee workers of the year, signalling that nearby nests are successfully producing young. Flowering slightly later than lungwort, leopard's bane (*Doronicum* spp.)** has lemon-yellow daisy-like flowers that are adored by bees. It's a bit old-fashioned and can be hard to find, but is worth seeking out.

Many wild thistles are thuggish weeds best kept out of the garden, but there are lots of garden-friendly thistles – and thistle lookalikes – that are loved by bees. Some are perennial, others biennial. Plume thistle (*Cirsium rivulare* 'Atropurpureum')*** has plum-coloured flowers that first appear in May and continue throughout June. It's a disappointingly short-lived plant and, being sterile, will not produce seedlings. If you want it to last, take root cuttings every couple of years. The cotton thistle, *Onopordum acanthium***, is a magnificent plant on an epic scale, with candelabras of flowering stems up to 3m high and tennis-ball flower heads that are adored by bees. It's biennial rather than perennial and seeds itself around rather freely, so needs some stern management. Cardoons (*Cynara cardunculus*)*** start life in spring as a modest fountain of shapely glaucous leaves but then grow enormous. Their huge purple punk-haired flowers are big enough to host a multitude of bees all at once. Globe thistles (*Echinops* spp.)*** have horribly prickly leaves but the blue pom-pom flower heads are so appealing to bees that they justify the occasional painful encounter. Sea hollies (*Eryngium* spp.)*** are related to carrots, but their spiky leaves and flowers earn them a spot in the thistle lookalike category. They come in a range of electric silvery-blue-green colours and when in flower will always have bees around them. Like the globe thistles, they like poor, well-drained soil and full sun, so are suited to our increasingly hot, dry summers.

Thistle-ish knapweeds are irresistible to bees. *Centaurea montana*** has lovely purple and mauve blooms with an exceptionally long flowering period from May to September. *C. macrocephala*** is more statuesque than the other species, with a yellow frizz of a flower that looks like something created by Jim Henson. Small solitary bees love to get lost among the anthers.

ABOVE Honey bees on cardoon.

Known as composites, anything shaped like a daisy is likely to attract custom. An open arrangement of petals around the outside and easily accessible florets in the middle simply shouts 'food' to passing bees. The ultimate in easy-to-care-for daisies is Mexican fleabane (*Erigeron karvinskiansus*)**, which once established will self-seed reliably, popping out of cracks and crevices in walls and patios with delightful puffs of small, pinky-white flowers. Next up in flower size are the asters (*Aster* spp. – though some have been reclassified as *Eurybia* or *Symphotrichum*)***. These tremendously useful plants form dense bushes of various shapes and sizes, but all produce masses of flowers that are visited by bees from about August to November. Two favourites of honey bees are *Aster* × *frikartii* 'Mönch' and *Aster* 'Little Carlow'. Other essential late-flowering daisies are the sneezeweeds (*Helenium* spp.)***, coneflowers (*Echinacea* spp.)***, black-eyed Susans (*Rudbeckia* spp.)*** and the perennial sunflowers (*Helianthus* spp.)***.

Many perennials produce flowers on spires, giving height to the display as well as convenient viewing opportunities as bees work their way up and down the spires. Biennial native foxgloves (*Digitalis purpurea*)**

are a good example, although their deep flowers mean they are used only by the bees with the longest tongues. Some of the ornamental foxgloves are perennial, with shorter, more open flowers that are accessible to a wider range of bees. Try *Digitalis* × *mertonensis* or *D. ferruginea*. Bugbane (*Actaea simplex*)***, doesn't sound like it would attract any kind of insect, but the long, bottle brush-like flower spikes are adored by many species of bees which can often be seen at work on the pinkish-white flowers. Bear's breeches (*Acanthus* spp.) have prehistoric-looking flower spikes that are very attractive to bumblebees, but beware – the flowers occasionally trap larger bees, which can sometimes be found dead inside them. The elegantly architectural flower spikes of Culver's root (*Veronicastrum virginicum*)*** are arranged like a candelabra on a slender stem. Each has thousands of tiny, bell-shaped flowers that are loved by honey bees and short-tongued bumblebees. The archetypal English herbaceous perennial, delphiniums seem to get more attention from slugs than from bees. A good alternative are the monkshoods, particularly *Aconitum* 'Bressingham Spire'**, which is enjoyed by long-tongued bumblebees but is toxic to animals and humans.

ABOVE The small scissor bee (*Chelostoma companularum*) is closely associated with campanulas.

Campanulas or bellflowers (*Campanula* spp.) come in a range of forms and colours. Choose which you like the most, but do have at least some. Members of this species are the favoured foraging plants for the UK's smallest solitary bee, the ant-sized *Chelostoma campanularum* – the clue is in the name. These diminutive bees are easily mistaken for flies as they buzz around campanulas. They are fun to watch when inside the flowers, scraping pollen off the anthers with the scissor-like actions of their legs. Milky bellflower (*Campanula lactiflora*) is particularly floriferous and will attract many bees.

There is quite a catalogue of salvias to choose from, and almost all are good for bees. The very long, tubular flowers of species like *Salvia fulgens*** are accessible from the front only by the long-tongued garden bumblebee (*Bombus hortorum*), but others will get to the nectar through holes nibbled in the base. One of the best salvias for honey bees and bumblebees is *Salvia × sylvestris*** and its various cultivars, which can be alive with bees for six weeks or more. Other equally recommended species are *S. nemorosa*, *S. verticillata*, and *S. yangii*, also known as perovskia or Russian sage. The

fashionable *Salvia* 'Amistad' can be very well worked by bees from May through to November.

What are frequently thought of as geraniums are often pelargoniums – the blousy, brightly coloured flowers often sold as bedding plants. Genuine geraniums have lovely little blooms and papery petals, often with delicate tracery on them. There are hundreds of species and cultivars, all of which are hardy, easy to grow, and flower profusely. All are ideal for a bee garden and handy for filling shady spaces under bigger plants. For sheer longevity choose the RHS Plant of the Centenary, *Geranium* 'Rozanne'***, which flowers for months on end without the need for deadheading.

The mauve flowers of lamb's ears (*Stachys byzantina*)*** are popular with many kinds of bees, but this plant's close association with one species in particular makes it essential for the bee garden. Male wool carder bees (*Anthidium manicatum*) stake out the plants and defend them territorially because they know that females will visit to harvest the silvery leaf hairs for their nests. If you have lamb's ears in your garden, you are very likely to attract wool carder bees. They also favour the biennial great mullein (*Verbascum thapus*), purple toadflax (*Linaria purpurea)* and, for something really unusual, try *Salvia argentea*, which has ruffled, down-covered leaves.

Sedum, also called stonecrop or ice plant (*Hylotelephium spectabile*), is like an aircraft carrier for insects, with butterflies, bumblebees and solitary bees continually landing and taking off. Flowering from about mid-August, it provides a good opportunity to watch bees up close as they scurry from floret to floret across the gently domed flowerheads. These plants can get a bit leggy, so give them a Chelsea chop in about May to encourage a neater, more compact plant with lots of flowers. The same treatment can be given to many other perennials, including asters, heleniums, *Nepeta*, *Rudbeckia* and *Echinops*.

Other recommended herbaceous perennials for bees:

- Alkanet (*Anchusa* spp.)***
- Bergamot/bee balm (*Monarda didyma*)**
- Bleeding heart/dicentra (*Lamprocapnos spectabilis*)**
- Caryopteris (*Caryopteris* spp.)**
- Columbine (*Aquilegia vulgaris*)**
- Dyer's camomile (*Anthemis tinctoria*)
- Euphorbia (*Euphorbia* spp.)*
- False indigo (*Baptisia australis*)**
- Geums (*Geum* spp.)**
- Gayfeather (*Liatris spicata*)***
- Giant scabious (*Cephalaria gigantea*)***
- Hollyhock (*Alcea rosea*)**
- Jacobs ladder (*Polemonium* spp.)**
- Japanese anemone (*Anenome* spp.)**
- Lupins (*Lupinus polyphylus*)*
- Ligularia (*Ligularia* spp.)
- Masterwort (*Astrantia spp.*)**
- Oriental poppy (*Papaver orientale*)**
- Obedient plant (*Physostegia virginiana*)***
- Penstemon (*Penstemon* spp.)*
- Peonies (*Paeonia* spp.)*
- Persicaria (*Persicaria* spp.)**
- Phlomis (*Phlomis* spp.)**
- Scabious (*Knautia* and *Scabiosa* spp.)***
- Selenum (*Selenum wallichianum*)**
- Siberian bugloss (*Brunnera macrophylla*)**
- Sidalcia (*Sidalcea malviflora*)**
- Speedwell (*Veronica* spp.)***
- Thrift (*Armeria maritima*)**
- Tickseed (*Coreopsis verticillata*)**
- Valerian (*Centranthus ruba*)**
- Virginia mallow (*Sida hemaphrodita*)**
- Verbena (*Verbena bonariensis, V. hastata, V. stricta*)***
- Yarrow (*Achillea* spp.)**

ABOVE Honey bee on sedum, a reliable late-season source of forage.

Some herbaceous perennials for bees:

LEFT TO RIGHT FROM TOP, THIS PAGE *Salvia nemorosa*, Perennial sunflowers (Helianthus), Veronica speedwell, Culver's root, Persicaria, Doronicum, Bergamot, *Aster × frikartii*, *Verbena bonariensis*, Lamb's ears, *Geum* 'Tangerine dream'.

OPPOSITE PAGE Knautia, Perovskia, Echinops; Eryngium, Helenium, Phlomis; Giant scabious, *Erigeron karvinskianus*, *Geranium* 'Rozanne'; Japanese anemone, Caryopteris, Echinacea.

Lawns, meadows and wild gardens

Lawns needn't be lifeless, but millions of gardens have at their heart a patch of grass that might as well be tarmac as far as bees are concerned, often with nothing more than inch-high bright green grass and the occasional unloved daisy – or worse still, a carpet made from unsustainable plastic. There are several approaches to creating a more floriferous lawn, depending on its size, the amount of work you want to put in and whether you want to retain the area as a recreational space. The benefit to bees and other wildlife can be enormous, but a patch of wildflowers can also be one of the most personally revivifying things you can create in your garden, providing somewhere for you to lie in the sun, breathe deeply and listen to the bees.

Flowering lawns

Most mature lawns already have some flowering plants growing in them; typically including clovers, plantains, dandelions, daisies and self-heal. Simply by altering your mowing regime these plants – normally decapitated while in bud – can burst into flower and provide resources for bees and other insects. After being pollinated the flowers will set seed, meaning that over several years they will multiply.

For low-growing colourful summer flowers, mow the lawn as usual until about mid-April, always removing the clippings. Then allow the flowering plants to grow and bloom. Once the first flush of flowers has finished, typically around the end of June, mow the lawn again. With the blade at its highest setting, you will deadhead the flowers and can encourage a second and maybe even a third flush. Many plants treated this way become prostrate, hugging the ground and flowering at much shorter heights than usual, creating a carpet of flowers. Once these flowers have set seed, you can return to

your normal mowing regime, producing a neat lawn for autumn and winter.

Alternatively, don't mow at all between late spring and late summer. An array of taller wildflowers can then grow, providing food, shelter and nesting sites for a much wider range of wildlife. The area will still tend to be dominated by grasses, some of which might occasionally be worked by honey bees for pollen and will also produce attractive seedheads that will feed birds. Alternatively, use both of the above methods, cutting some of the lawn after flowering and allowing other areas to grow. A variety of heights can bring a sculptural quality to the garden.

People worry that a lawn left to grow will look untidy, or earn them a reputation as an inattentive gardener. One solution is to mow a pathway through or around the edge of the lawn. Doing so regularly neatens things up, shows the garden is cared for and creates somewhere to walk without fear of disturbing flowers and wildlife. You can even mow a different-shaped path every year to give your garden strolls some variety, or mow larger areas to be designated for recreational use. A few patches of very short grass in sunny places will help to encourage some species of mining bee to nest in your lawn.

Letting an established lawn regenerate in this way won't always produce the variety of flowers you might have hoped for, in which case they can be introduced. The easiest way is with individually grown perennial plug plants. These can be bought from specialist nurseries or grown at home from seed planted in modular trays. Providing a native wildflower is not protected (very few are) and you don't uproot it, it's fine to harvest seeds sparingly from plants found growing in the wild – although you should seek permission from the landowner. Plants from locally sourced seeds are more likely to do well in your garden. Plant out plugs in the autumn after the grass has had a final cut, allowing them to establish themselves before the following summer.

LEFT A well-established native perennial meadow.

Recommended wildflower species for short grass:

- Black medick (*Medicago lupulina*)*
- Bird's-foot trefoil (*Lotus corniculatus*)***
- Cowslip (*Primula veris*)*
- Dove's foot cranesbill (*Geranium molle*)*
- Dandelion (*Taraxacum officinale*)***
- Eyebright (*Euphrasia* spp.)*
- Ground ivy (*Glechoma hederacea*)**
- Lesser trefoil (*Trifolium dubium*)**
- Primrose (*Primula vulgaris*)**
- Red clover (*Trifolium pratense*)***
- Self-heal (*Prunella vulgaris*)**
- White clover (*Trifolium repens*)***
- Wild thyme (*Thymus polytrichus*)***

THIS PAGE Primrose; Horseshoe vetch; Dandelion, Doves's foot geranium, White clover; a flowering lawn and common carder bee (*Bombus pascuorum*).

OPPOSITE PAGE Bird's-foot trefoil and the beautiful but scarce gold-fringed mason bee (*Osmia aurulenta*); Ground ivy; Selfheal, Black medick.

Once these plants flower and set seed, your lawn will begin to become a self-sustaining mini meadow.

For areas of longer grass, a key plant is yellow rattle (*Rhinanthus minor*). Sometimes called the meadow maker, this annual flower semi-parasitises the roots of grasses, preventing them from outcompeting and suppressing wildflowers. Yellow rattle grows best from freshly produced seed planted in the autumn, either in small pots for later transplanting, or planted directly into prepared patches of bare soil. After flowering, allow it to set seed before you mow the grass. Leave the clippings for a week or so before raking them up so the seeds fall to the ground. Red bartsia (*Odontites verna*) performs the same function and might even attract the red bartsia solitary bee (*Melitta tricincta*). Both of these plants restrain thuggish grasses by depriving them of nutrients, so their work shouldn't be undone by applying any form of fertiliser to a wildflower lawn. Indeed, most wildflowers grow best in poor, relatively unfertile soil. For this reason, always remove mown clippings from any flowering lawn or meadow, to prevent them from composting down and enriching the soil.

Finally, a word about LYJs – 'little yellow jobs'. There are a host of very similar-looking small yellow flowers that are generally of use to bees, particularly small solitary bees. They include hawkbits, hawksbeards, hawkweeds, catsear and nipplewort. If you are a keen botanist it must be fun to decipher which is which. For the rest of us it's just useful to know that, if it's small and yellow, it's probably useful to pollinators.

BELOW Yellow rattle helps flowering meadows to thrive.

Recommended wildflower species for longer grass:

- Common knapweed and greater knapweed (*Centaurea nigra, C. scabiosa*)***
- Devil's-bit scabious (*Succisa pratensis*)***
- Field scabious (*Knautia arvensis*)***
- Mallow (*Malva* spp.)**
- Meadow clary (*Salvia pratensis*)***
- Meadow cranesbill and dusky cranesbill (*Geranium pratense, G. phaeum*)***
- Ox-eye daisy (*Leucanthemum vulgare*)**
- Red bartsia (*Odontites verna*)**
- Ragged robin (*Lychnis flos-cuculi*)**
- Wild marjoram (*Origanum vulgare*)***
- Yellow rattle (*Rhinanthus minor*)**
- Yarrow (*Achillea millefolium*)**

LEFT TO RIGHT FROM TOP Kidney vetch; Meadow cranesbill; Mallow, Ragged robin, Ox-eye daisy; Wild oregano, Field Scabious, Meadow Clary, Common knapweed.

ABOVE The short-lived firework display of an annual meadow.

Meadows

The definition of a meadow is an area of perennial grasses and wildflowers that is maintained by an annual hay cut. True wildflower meadows take decades, even centuries, to mature and achieve an exquisite equilibrium of many interdependent species of plants and animals. However, for most of us, any fairly large patch of garden filled with grasses and colourful flowers can be classed as a wildflower meadow – even when many of the flowers aren't very wild or indeed native. Wildflower meadows can attract an enormous variety of bees and other insects and there can be few more glorious places to be, late on a summer's day, the sinking sun making insects sparkle as they flit and fly between flowers to a soundtrack of contented humming, thrumming and chirruping.

Broadly, there are two kinds of garden meadow. Perennial meadows have a mix of grasses and flowering plants that return year after year. There is a choice between something that closely resembles a traditional wildflower meadow with a naturalistic appearance and native plants, or a more contemporary mix of species, designed to have a particular visual impact or be especially useful to pollinators.

Annual meadows have their roots in old-fashioned arable farming techniques that allowed for the flowering of various pretty cornfield flowers. These grew afresh each year from seeds produced the summer before. Modern annual meadows often contain many species

not naturally found together, and tend to be bright, bold and impactful.

Most meadows are grown from seed. In the case of a perennial meadow this should only have to be done once, but an annual meadow will need to be reseeded each year. Choose your seed mix carefully, according to your desired effect and your soil type and location. Specialist seed merchants can advise on this. If your topsoil is very rich it can be worth removing the top layer to reduce fertility before planting. Planting should be done in late summer/early autumn or in spring, scattering the seeds on bare, prepared soil before lightly raking them in, ideally rolling them to help to lock in moisture. Water if conditions are dry and continue to do so until established. Although it can be difficult to tell weeds from seedlings, remove any identifiable weeds as they grow. It is not necessary to add any fertiliser.

For faster, more reliable but costlier results, ready-grown wildflower turf can be used. Your chosen mix of plants is supplied in rolls that are laid like a carpet on prepared soil for an almost instant effect. The growing meadow still has to be maintained, but this method gets things off to a great start.

Perennial meadows

Perennial meadows can contain a mix of perennial native and perhaps some non-native wild species that, when established, regrow each year. They can include some annual flower species too, as well as various non-invasive grass species. Eventually, they should be self-sustaining – the plants replenishing themselves by spreading or self-seeding. They are visually stunning in a delightfully understated way, with a subtle mix of green and golden grasses scattered with jewel-like flowers that invite closer inspection for full appreciation. Yellow rattle and red bartsia are important components of any perennial meadow, for the reasons mentioned earlier.

Even well-established perennial meadows must be managed on a regular basis, mainly to prevent grasses and unwanted weeds from suppressing the wildflowers. Control broadleaved weeds by removing the flower heads so that they cannot set seed, and dig them out where necessary. Mowing regimes can vary depending on the site and what you want to achieve. An early spring cut will remove recently grown grass and give the emerging wildflower plants a better chance to flourish. Mow again in late summer after flowering and once seed has set, raking up the cuttings so that the seeds fall to the ground. Collect the seed of species you want more of, and grow them as plugs to fill gaps the following year. Experimenting with how and when you cut can produce a variety of flowering times and meadow styles. You may need to use a strimmer for the autumn cut, followed by a mower. Scything is a more physically and perhaps spiritually rewarding way to cut your meadow, and undoubtedly more environmentally friendly.

Annual meadows

Annual meadows are the horticultural equivalent of a firework display, bursting with effervescent colour and rich in nectar and pollen for about three months, before fizzling out with barely a trace. They can be planted in any sunny area, replacing a whole lawn or just filling a gap in a border. They are a great way to liven up a small front garden or the edge of a path. You can even create a mini meadow in a pot or a grow bag. For an old-fashioned look use native cornfield species, but almost any annual flowers will work. Many meadow seed mixes are available, giving a variety of styles, heights and colours, meaning you can change the look of your meadow each year to find out what works best for you and the bees.

Annual meadows are fairly easy to maintain, though some plants benefit from regular deadheading and sometimes a bit of staking. Although they will self-seed to a degree, they really only remain colourful and vigorous if replanted each year – removing the old plants in autumn, preparing the ground and re-seeding immediately or in late spring. Unlike perennial species, annuals are fine in rich soil, rocketing away before grasses have a chance to take hold.

BELOW Large-headed resin bee (*Heriades truncorum*) on corn marigold.

Recommended annual meadow species:

- Dahlias (*Dahlia* spp.)***
- Borage (*Borago officinalis*)***
- California poppy (*Eschscholzia californica*)**
- Corncockle (*Agrostemma githago*)*
- Cornflower (*Centaurea cyanus*)**
- Corn marigold (*Chrysanthemum segetum*)*
- Cosmos (*Cosmos bipinnatus*)***
- Dill (*Anethum graveolens*)*
- Field poppy (*Papaver rhoeas*)*
- Honesty (*Lunaria annua*)**
- Phacelia (*Phacelia tanacetifolia*)***
- Pot marigold (*Calendula officinalis*)**
- Safflower (*Carthamus tinctorius*)***
- Serradella (*Ornithopus sativus*)***
- Sunflower – dwarf varieties (*Helianthus annuus*)***
- Toadflax (*Linaria* spp.)*

LEFT TO RIGHT FROM TOP Pot marigold; Phacelia; Cornflower, Cosmos; California poppy, Toadflax.

*Recommended species for perennial meadows
include all of those suggested earlier as well as:*

- Betony (*Betonica officinalis*)***
- Creeping thistle (*Cirsium arvense*)***
- Common fleabane (*Pulicaria dysenterica*)**
- Dyer's greenweed (*Genista tinctoria*)*
- Goldenrod (*Solidago* spp.)***
- Grass vetchling and meadow vetchling (*Lathyrus nissolia, L. pratensis*)***
- Great mullein (*Verbascum thapsus*)*
- Harebell (*Campanula rotundifolia*)**
- Hemp agrimony (*Eupatorium cannabinum*)**
- Vetches (*Hippocrepis* spp.)***
- Meadowsweet (*Filipendula ulmaria*)**
- Meadow rue (*Thalictrum flavum*)**
- Melilot, white and ribbed (*Melilotus albus, M. officinalis*)***
- Red bartsia (*Odontites verna*)**
- Red campion (*Silene dioica*)*
- Sainfoin (*Onobrychis viciifolia*)***
- Viper's bugloss (*Echium vulgare*)***
- Wild carrot (*Daucus carota*)*
- Yarrow (*Achillea millefolium*)**
- Yellow rattle (*Rhinanthus minor*)**

THIS PAGE Tufted vetch; Goldenrod; Hemp agrimony; a species-rich perennial meadow.

OPPOSITE PAGE, LEFT TO RIGHT FROM TOP Ribbed melilot, Creeping buttercup; Creeping thistle; Meadowsweet, Red campion; Red bartsia, Betony, Sainfoin.

Wild gardens

After millions of years of co-evolution, many flowers and bees have developed a perfect, co-dependent relationship. In some cases, both flower and bee are physically suited to one another, fitting together like a lock and its key. However, many wild plants have been selectively bred to produce garden versions that are more colourful, shapely, floriferous or long-flowering. For some species, this has broken the age-old relationship between flower and bee; perhaps the shape or structure of the flower no longer easily allows access to pollen and nectar, or there may be none to be found, if the plant's ability to reproduce naturally has been bred out of it. One way to make sure your garden is full of plants that bees will love is to grow plenty of native, wild species of plants. Unchanged by horticultural tinkering, these still offer bees a sure source of nectar and pollen.

Growing native wild plants in your garden is not akin to rewilding, in which landscapes are left to regenerate without human intervention. Leaving the average garden untended would result in an unruly tangle in which a small number of species would dominate. In fact, growing native plants in your garden demands just as much thought and attention as any other kind of gardening, with various decisions to be made about the approach you take. For example, what is your definition of wild or native? Which plants should be allowed to spread and which should be strictly controlled? What is allowed to grow in beds or borders and what should be confined to a wild corner, and what is your boundary between a wildflower and a weed?

Everyone has different thresholds when it comes to what they consider a weed. Most people would agree that the native stinking hellebore (*Helleborus foetidus*)** is an attractive bee-friendly native wildflower to be welcomed in any garden. But while I am delighted by dandelions in my lawn, for some that is a step too far. Some native plants that are good for bees really are best avoided because they are more trouble than they are worth. Hedge bindweed (*Calystegia sepium*)*, for example, is loved by bumblebees, but is so rampant that surely no gardener would introduce it willingly. Hogweeds (*Heracleum* spp.)*** can be beautiful and statuesque additions to the wild garden and are adored by pollinators, but their sap can cause severe blisters to the skin and they should be treated with extreme caution. Common ragwort (*Senecio jacobaea*)*** is poisonous to

ABOVE This Gold Medal-winning garden at the 2022 Chelsea Flower Show demonstrated how a version of the wild might be created in a small garden – beaver dam optional.

ABOVE Two of the most important native plants; ivy (left, with honey bee) and ragwort (right, with a male patchwork leafcutter bee).

livestock, yet is one of the best of all plants for wildlife. More than 200 species of invertebrates have been recorded living on it, and 35 species of insects are wholly dependent upon it. It's worth having if only to see the striking caterpillars of the cinnabar moth, for whom it is their only food. It is not illegal to let ragwort grow in your garden, but you must not cause it to grow in the wild or spread to someone else's land. Does this mean you should welcome it if it arrives, and will you have a plan that stops it from spreading? The most important rule of thumb when it comes to wild plants is perhaps horticulture's most repeated mantra – right plant, right place.

Many native species can be bought as seeds or plants but can also be found in hedgerows and woodlands, so start by looking for those that already grow well in your area and collect seed or take cuttings of those you like, or see bees using. It's especially worth looking for any that are known to have a relationship with a particular species of bee; if a plant is growing nearby, there is a chance that there might be a local population of bees that depend on it. The most likely specialist to visit your garden is the ivy bee (*Colletes hederae*), which almost exclusively uses the flowers of ivy. Native ivy (*Hedera helix*)*** is among the most valuable UK plants for insects, including bumblebees, honey bees, butterflies, wasps and hoverflies, all of which use it primarily for nectar as winter approaches. Many creatures overwinter among its leaves and tendrils, and the berries are an important food for

birds. It is the older, bushier ivy that flowers, so try to let some thicken up and bloom. Tales that it harms trees and brickwork are generally exaggerated. If native ivy feels a little too mundane for your garden, various ornamental species are available. Persian ivy (*Hedera colchica*) is a little more glamorous while still producing flowers that are useful to wildlife. The following native plants are also worth planting to encourage their associated specialist bees should they be present in your area:

- Buttercups (*Ranunculus* spp.) – large scissor bee (*Chelostoma florisomne*)
- Campanulas (*Campanula rotundifolia, C. glomerata, C. lactiflora C. trachelium*) – bellflower blunthorn bee (*Melitta haemorrhoidalis*) and small scissor bee (*Chelostoma campanularum*)
- Mignonette/weld (*Reseda* spp.) – large yellow-face bee (*Hylaeus signatus*)
- Red bartsia (*Odontites verna*) – red bartsia bee (*Melitta tricincta*)
- Scabiouses (*Knautia, Scabiosa, Succisa* spp.) – small scabious mining bee (*Andrena marginata*) and large scabious mining bee (*Andrena hattorfiana*)
- White bryony (*Bryonia dioica*) – bryony mining bee (*Andrena florea*)
- Yellow loosestrife (*Lysimachia vulgaris*) – yellow-loosestrife bee (*Macropis europaea*).

Other wild plants to attract bees to your garden or wild area:

- Alexanders (*Smyrnium olusatrum*)**
- Bastard balm (*Melittis melissophyllum*)***
- Betony (*Betonica officinalis*)**
- Black horehound (*Ballota nigra*)***
- Brambles (*Rubus* spp.)***
- Bugle (*Ajuga reptans*)**
- Burdocks (*Arctium* spp.)**
- Coltsfoot (*Tussilago farfara*)**
- Common restharrow (*Ononis repens*)**
- Cranesbill geraniums (*Geranium* spp.)***
- Dandelion (*Taraxacum officinale*)***
- Deadnettles (*Lamium* spp.)***
- Dog rose (*Rosa canina*)**
- Foxglove (*Digitalis purpurea*)***
- Figworts (*Scrophularia* spp.)**
- Gorse (*Ulex europaeus*)**
- Green alkanet (*Pentaglottis sempervirens*)**
- Ground ivy (*Glechoma hederacea*)**
- Houndstongue (*Cynoglossum officinale*)***
- Mulleins (*Verbascum* spp.)**
- Primrose (*Primula vulgaris*)**
- Rosebay willowherb (*Chamaenerion angustifolium*)***
- Tansy (*Tanacetum vulgare*)***
- Teasel (*Dipsacus fullonum*)***
- Thistles (*Cirsium* spp.)***
- Toadflax (*Linaria vulgaris*)**
- Traveller's joy (*Clematis vitalba*)***
- Valerian (*Centranthus* spp.)**
- Vetches (*Vicia* spp.)***
- Weld (*Reseda luteola*)**
- White horehound (*Marrubium vulgare*)***
- Woundworts (*Stachys* spp.)***
- Wood sage (*Teucrium scorodonia*)***
- Yellow archangel (*Lamium galeobdolon*)***

LEFT TO RIGHT FROM TOP Yellow archangel, Rosebay willowherb, Gorse; Common vetch, Alexanders, Hedge woundwort; Traveller's joy, Lesser burdock, White dead-nettle; Green alkanet, Weld, Wild geranium.

Pools and damp gardens

The best single thing you can do to support wildlife of all kinds in your garden is to make a pond. Anything with water and a few plants – even it is only a bucket or an old sink – will improve biodiversity. A pond of any size, set in the ground with gently sloping sides, is the best option, allowing a multitude of creatures to access the water and surrounding habitat. It's astonishing how within hours of water being added, pond skaters and damselflies can arrive, with frogs and newts often not far behind. A whole web of water-dependent wildlife will usually establish itself within a matter of months, with very little help from you. Among other insects, ponds will attract honey bees, which will visit to collect water, particularly in early spring. For this reason, try to have areas of shallow or boggy water that receive winter sunshine to warm the water.

Ponds can be the beating heart of a wildlife garden, but a rich environment can develop on their peripheries. Marginal plants grown in the shallow water and boggy areas at the edge of a pond will provide flowers for pollinators, and stems and roots where other creatures can live and breed. Where the ground is not permanently wet but just moist, a host of damp-loving plants can be grown, adding not only colour and texture but also valuable sources of nectar and pollen for bees.

Pond plants for bees

A small number of flowering aquatic plants useful to bees can be grown in the deeper parts of ponds. Sadly, water lilies don't seem to interest bees very much, being pollinated mostly by beetles – though lily pads are a useful drinking platform for bees. The fringed water lily (*Nymphoides peltata*)*, whose small, yellow flowers last just one day, is sometimes visited by honey bees,

probably only for pollen. Water hawthorn (*Aponogeton distachyos*)* is a good lily alternative. Growing from fleshy roots in deep water, it produces long, semi-evergreen leaves and usually flowers in two flushes, in early and late summer. The creamy-coloured, vanilla-scented flowers attract various kinds of bees. Water crowfoot (*Ranunculus aquatilis*)* can be grown as an oxygenator, and produces rafts of white and yellow daisy-like flowers that are visited by pollinators.

Marginal plants

A wide variety of flowering plants attractive to bees can be grown around the margins of ponds. Many are not too fussy about the depth of water they grow in, but some can be quite particular, so check with your supplier about the best planting depths.

Marsh cinquefoil (*Potentilla palustris*)*** should be at the top of the list of marginals in any beekeeper's pond. Also known as purple marshwort or bog strawberry, this low-growing native perennial produces a carpet of narrow, silvery-green floating leaves that provide excellent cover for all kinds of pond life. Dark maroon flowers, looking somewhat like miniature wild roses with a strawberry in the centre, are produced profusely in June and July and are highly attractive to bees. This plant is not as

LEFT A pond can provide a source of water for bees as well as pollen- and nectar-rich flowers.

RIGHT Marsh cinquefoil.

common in garden ponds as it should be and is usually only available from specialist water plant nurseries, although in the Orkney islands it is considered a major source of nectar for honey bees. Another rarity worth seeking out is American water willow (*Justicia americana*)**. This grows airy stems with willow-like leaves, topped by exquisite white and lavender-pink orchid-like flowers. It attracts honey bees and bumblebees. Although native, bogbean (*Menyanthes trifoliata*)** has exotic-looking, white and pink, star-shaped flowers in late spring. Loved by honey bees, bogbean is also an important food for the enormous caterpillars of the tropical-looking elephant hawk moth. For early colour, plant the native marsh marigold or kingcup (*Caltha palustris*)*, usually the first pond plant of the year to flower. On mild spring days it can attract honey bees and some species of solitary bee. Several cultivars are available, but the wild form is the most spectacular. Depending on available space, one of the various pickerel weeds is a must. Most commonly grown are the compact *Pontederia cordata*** and its various cultivars. All produce tall, upright stems and elegant, lance-shaped leaves. The flowers are hyacinth-like spires of blue, pink, purple or white. In larger ponds, *P. dilatata*** and *P. lanceolata*** are much grander plants, reaching up to 1.5m (5ft). All pickerels attract bumblebees and honey bees and are also favoured by the emerging nymphs of dragonflies and damselflies. Another dragonfly favourite is the narrow-leaved water plantain (*Alisma lanceolatum*)* which has tiny white flowers that are very attractive to bumblebees, which bob about precariously when they land on them. This plant's airy flower stems add lightness and delicacy to the water's edge.

BELOW Common carder bee on water willow.

Other recommended marginal plants:

- Amphibious bistort (*Persicaria amphibia*)*
- Common monkey flower (*Mimulus guttatus*)*
- Creeping jenny (*Lysimachia nummularia*)*
- European water plantain (*Alisma plantago-aquatica*)**
- Flowering rush (*Butomus umbellatus*)*
- Lizard's tail (*Saururus cernuus*)***
- Water forget-me-not (*Myosotis palustris*)*
- Water mint (*Mentha aquatica*)***
- Watercress (*Nasturtium officinale*)*
- Water obedient plant (*Physostegia leptophylla*)**
- Irises (*Iris pseudacorus*, *I. sibirica*)*

BELOW Water mint (top); Kingcup (bottom).

OPPOSITE PAGE, LEFT TO RIGHT FROM TOP A buff-tailed bumblebee getting stuck into a yellow flag iris; Pickerel weed, Bogbean; Narrow-leaved water plantain, Common monkey flower.

ABOVE A green-eyed flower bee (*Anthophora bimaculata*) on damp-loving Hooker's inula.

The damp garden

As climate change dials up summer temperatures, damp gardens and the flowering plants that can grow in them are likely to become increasingly important to pollinators. The perennial plants listed below all grow well in damp but free-draining soil. Most will tolerate seasonal flooding but, unlike the marginal species mentioned earlier, dislike having wet feet all the time. If your ground is not damp naturally, it is easy to make a bog garden using waterproof membrane or pond liner.

Purple loosestrife (*Lythrum salicaria*)*** is an elegant native plant with spires of purple-pink flowers that are rich in nectar and pollen and much visited by bees, butterflies and moths. It is a vigorous self-seeder and can produce dense clumps that can be invasive, though rarely in a garden situation. Yellow loosestrife (*Lysimachia vulgaris*)** is an unrelated but similarly vigorous native plant. Its bright yellow flower spikes, up to 120cm high, are attractive to various bees but notably the yellow-loosestrife bee (*Macropis europaea*), a solitary mining bee that waterproofs its tunnels with oil harvested from yellow loosestrife flowers. Gooseneck loosestrife (*Lysimachia clethroides*)*** has clusters of white flowers on charmingly curved stems, and is keenly worked by honey bees, bumblebees and solitary bees of various species. Hooker's inula (*Inula hookeri*)*** will bring a high-impact splash of sunshine-yellow even to shadier spots. Its cheery, frilly sunflower-like flowers twist open from fluffy buds and attract masses of bees and butterflies. The damp-loving varieties of cardinal flower (*Lobelia cardinalis*)** are classic waterside flowers that bring an exotic feel to the garden. There are many species but *L. siphilitica* and the purple-leaved *L. cardinalis* are superb bee plants. Although far less showy, the low-growing water avens (*Geum rivale*)*** more than compensates with its appeal to bumblebees and solitary bees such as the hairy-footed flower bee. There are various colourful cultivars, but the subtle pink and yellow nodding bells of the wild UK native are perhaps the prettiest of all. At the other end of the height scale, the towering, dusky purple flowerheads of Joe Pye weed (*Eupatorium purpureum*)** can provide welcome late-season nectar for bees of all kinds. In the USA it's considered an important honey plant. Similarly statuesque but more elegant is the sweetly perfumed common valerian (*Valeriana officinalis*)***, loved by bees and hoverflies. For something beautifully sinister in gardens not frequented by children, any of the dropworts will

ABOVE Damp areas surrounding even very small ponds can be rich habitats.

invite a host of pollinators – but all are highly poisonous to humans. Tubular water dropwort (*Oenanthe fistulosa*)** puts on a lovely display of white pin-cushion flowers that attract bees and butterflies from mid- to late-summer.

Himalayan balsam (*Impatiens glandulifera*)*** is one of the most controversial plants in the UK. Introduced by the Victorians as a garden exotic, it now lines thousands of miles of river bank, clogs streams and marshes and can cause riverbank erosion to the detriment of native wildlife. Balsam bashing events are held each year to somewhat optimistically pull up the easily dislodged plant before it sets seed. However, it is an exemplary bee plant and provides nectar for honey bees and bumblebees late in the season when little else is available. Whatever else you think about it, as a beekeeper you are lucky if it grows nearby. It can provide plenty of honey for your bees to overwinter on and possibly even enough for you to harvest. It is not illegal to grow Himalayan balsam in your garden, but it is illegal to allow it to spread to other property or to the wild. If you want to grow something similar but less rampant, try *Impatiens balfourii***.

Other recommended damp-loving plants:

- Angel's fishing rod (*Dierama* spp.)**
- Bog sage (*Salvia uliginosa*)***
- Bugbane (*Actaea simplex*)***
- Bugle (*Ajuga reptans*)*
- Butterbur (*Petasites hybridus*)***
- Camassia (*Camassia leichtlinii*)*
- Canadian burnett (*Sanguisorba canadensis*)**
- Culver's root (*Veronicastrum virginicum*)***
- Devil's-bit scabious (*Succisa pratensis*)***
- Globeflower (*Trollius europaeu*)*
- Greater bird's-foot trefoil (*Lotus pedunculatus*)***
- Gypsywort (*Lycopus europaeus*)**
- Himalayan honeysuckle (*Leycesteria formosa*)**
- Hostas (*Hosta* spp.)*
- Lady's smock (*Cardamine pratensis*)**
- Leopard plant (*Ligularia dentata*)**
- Marsh woundwort (*Stachys palustris*)***
- Meadowsweet (*Filipendula ulmaria*)**
- Narrow-headed leopard plant (*Ligularia stenocephala*)**
- Persicaria (*Persicaria amplexicaulis*)***
- Ragged robin (*Lychnis flos-cuculi*)**
- Teasel (*Dipsacus fullonum*)***

LEFT TO RIGHT FROM TOP:

THIS PAGE Bugle; Himalayan balsam; Teasel; Globeflower, Marsh woundwort, Water avens.

OPPOSITE PAGE Yellow loosestrife, Purple loosestrife; Common valerian; Himalayan honeysuckle, Leopard plant; Gooseneck loosestrife.

Recommended damp-loving trees and shrubs,
see next chapter for more details:

- Alder (*Alnus glutinosa*)**
- Alder buckthorn (*Frangula alnus*)**
- Black chokeberry (*Aronia melanocarpa*)*
- Button bush (*Cephalanthus occidentalis*)**
- Leatherwood (*Eucryphia lucida*)**
- Medlar (*Mespilus germanica*)**
- Goat willow (*Salix caprea*)***
- Quince (*Cydonia oblonga*)**

Hedges, shrubs and trees

Hedges, shrubs and trees give a garden structure, shape, volume and shelter. Second only to a pond, a good hedge is among the best things you can provide for garden wildlife. Birds nest and shelter in them, amphibians live and hibernate under them, their flowers benefit bees, and their berries feed birds. They provide corridors for wildlife to move between various habitats, and the dry, leafy area at the foot of a hedge can be a good place for bumblebees to hibernate or nest.

As well as the forage value of their flowers, hedges and shrubs are invaluable as windbreaks in the bee garden. They can buffer winds that make it difficult for honey bees to land and take off from their hives, and they shelter colonies from chilling winter draughts. A tall hedge not far from a hive will force bees to fly upwards, taking them up above head height of you and your neighbours. Reducing wind flow also makes it easier for bees, especially the smaller solitary species, to navigate around the garden and land on flowers that might otherwise be swaying in a breeze.

Hedge plants

The best hedge for bees and other wildlife is one that comprises a selection of flowering native species. Many flower in spring, providing pollen and nectar at a particularly crucial time, and later producing berries and hips for birds and small mammals. Most flower on older, woody growth, so the best time to trim a bee hedge is after flowering, usually in early summer, giving time for new shoots to grow and mature before flowering again the following year. The danger here is that you might disturb nesting birds, so be cautious – plus of

course removed flowerheads will mean fewer berries or hips. Better still, trim only every other year, allowing at least a full year of uninhibited growth, flowering and fruiting between cuts. If you have more than one hedge, alternate cutting years, giving just a light trim in late winter if you want a sharper edge. Hedging plants

LEFT Flowering in January, this winter cherry provides an early source of pollen and nectar.

RIGHT Growing wild in woods and hedgerows, hazel (top) and goat willow (bottom) are often the first major source of food in early spring.

can be bought individually, but most suppliers sell selections designed for birds, which can also be good for bees. They are usually available as very reasonably priced bare-rooted bundles for planting in winter and spring. Many of the plants listed below can be grown as specimen shrubs or trees, but will respond well to being planted and maintained as a hedge.

ABOVE Hawthorn, also called mayflower, is used by honey bees, bumblebees and early-emerging solitary bees. It can sometimes produce a crop of delicious honey.

BELOW Field maple, another reliable early-flowering tree.

Recommended flowering native hedge plants:

- Alder buckthorn (*Rhamnus frangula*)***
- Blackthorn (*Prunus spinosa*)**
- Bird cherry (*Prunus padus*)***
- Crab apple (*Malus sylvestris*)***
- Dog rose (*Rosa canina*)**
- Field maple (*Acer campestre*)**
- Goat willow (*Salix caprea*)***
- Hawthorn (*Crataegus monogyna*)***
- Hazel (*Corylus avellana*)**
- Holly (*Ilex aquifolium*)**
- Spindle (*Euonymus europaeus*)**
- Wild cherry (*Prunus avium*)***
- Wild privet (*Ligustrum vulgare*)***

Ornamental shrubs

Ornamental flowering shrubs can be grown as specimens, giving shape and structure to a border, but many make good hedge plants too. Hardy fuchsias are very wind-tolerant, with *Fuchsia magellanica**** often being grown near the south-west coast as a hedge to protect early flower and vegetable crops. It has naturalised in some places, and can be a valuable source of honey. Escallonia (*Escallonia* spp.)** also grows well in coastal areas. Its trumpet-shaped flowers, produced from about June until the first frosts, are adored by bees, which enter by the front or use holes nibbled in the base of the flower. With their wickedly sharp spines, the barberries make impenetrable hedges with flowers loved by bees, particularly short-tongued bumblebees. The golden barberry (*Berberis* × *stenophylla*)** is the one most used for hedging, although others, including Darwin's barberry (*B. darwinii*)** and the wintergreen barberry (*B. julianae*)**, are excellent too. Cherry laurel (*Prunus laurocerasus*)* can be grown as a shrub or a hedge and, though rarely recommended as a bee plant, does have some interesting benefits. In early spring its glossy leaves make a good sun trap, and I often see newly emerged solitary bees basking on them, especially the orange-tailed mining bee (*Andrena haemorrhoa*), the ashy mining bee (*A. cineraria*) and the tawny mining bee (*A. fulva*). The short-lived candelabras of white flowers are moderately attractive to bees in spring, but in early summer the young leaves develop extrafloral nectaries on their undersides. These can be very attractive to honey bees and the occasional wasp, particularly in years when there is a pronounced June gap. The almondy smell that comes from laurel is hydrogen cyanide, and in fact the crushed leaves were once used by entomologists to kill their specimens, though that doesn't seem to stop bees and wasps visiting the plant.

Tightly clipped, formal hedging might not be thought of as good for bees, but can sometimes be useful. Box (*Buxus sempervirens*) has modest yellow-green flowers in spring that are sometimes eagerly worked by honey bees. With box blight and box moth caterpillars creating problems, try instead either shrubby honeysuckle (*Lonicera nitida*) or box-leaved honeysuckle (*L. pileata*), both dense, fast-growing evergreens that in spring and early summer can be smothered in small flowers and appreciative bees. Flowering earlier, and

ABOVE Bees go wild for flowering *Cotoneaster horizontalis*.

with less dense foliage, sweet box (*Sarcococca confusa*) draws in honey bees with its heavy, honeyed scent. One of the best-known bee shrubs must be the wall cotoneaster (*Cotoneaster horizontalis*)***, which makes a low, narrow hedge when grown against a fence or a wall. Even those not particularly attuned to the comings and goings of bees often seem to notice the spectacle of bees mobbing the unshowy flowers of this plant. Cotoneasters form a large group and all are attractive to bees though some, including the wall cotoneaster, are counted as invasive UK non-native species that are fine to plant in your garden but should not be allowed to grow in the wild.

No garden should be without roses, but many are of little use to bees because of their thousands of years of selective breeding. The best kinds for bees are species roses which are wild, or close to wild, with an open structure, visible anthers and a rich scent. Bees of various kinds will visit them, although seemingly only for pollen. Some species of roses, typically those with very visible yellow stamens such as *Rosa moyesii* and *R. rugosa*, are buzz pollinated. This method, also called sonication, requires a bumblebee to wrap her legs around the stamens and vibrate her flight muscles. It is quite charming to sit under a rose bush on a summer's day and listen to the collective high-pitched buzzing of sonicating bumblebees, sounding like a miniature kazoo orchestra. Rose leaves are a favourite target for leafcutter bees, which leave cookie cutter-style holes in them after removing pieces for their nests. They do no harm and rose bushes should wear their holes as a badge of honour. If you wish to grow roses for bees, the best option is to visit a specialist rose nursery or order an illustrated catalogue, and look for single or semi-double-flowered species roses and their cultivars with the colour and habit you require. Recommended commonly available species roses include *R. canina*, *R. gallica*, *R. glauca*, *R. moschata*, *R. moyessi*, *R. multiflora*, *R. rubiginosa*, *R. rugosa* and *R. spinosissima*.

Other recommended flowering shrubs for bees:

- Abelia (*Abelia* spp.)**
- Aralia (*Aralia* spp.)***
- Bastard senna (*Coronilla valentina*)***
- Broom (*Cytisus* spp.)**
- Bottlebrush (*Callistemon* spp.)
- Buddleia (*Buddleja* spp.)**
- Californian lilac (*Ceanothus* spp.)**
- Callicarpa (*Callicarpa bodinieri*)**
- Camellias (*Camellia* spp.)*
- Chaste tree (*Vitex agnus-castus*)***
- Cistus (*Cistus* spp.) **
- Deutzia (*Deutzia scabra*)**
- Eucryphia (*Eucryphia* spp.)**
- Firethorn (*Pyracantha coccinea*)***
- Flowering currant (*Ribes sanguineum*)*
- Hebe (*Hebe* spp.)**
- Hop tree (*Ptelea trifoliata*)***
- Mallow (*Malva* spp.)**
- Mock orange (*Philadelphus* spp.)**
- Osmanthus (*Osmanthus* spp., especially × *burkwoodii*)***
- Pieris (*Pieris* spp.)**
- Rhododendron (*Rhododendron* spp.)**
- Rock rose (*Helianthemum* spp.)**
- Snowberry and coralberry (*Symphoricarpos* spp.)***
- Spiny oleaster (*Elaeagnus* spp.)***
- St John's wort (*Hypericum* spp.)**
- Tamarisk (*Tamarix* spp.)*
- Tree heath (*Erica arborea*)**
- Weigela (*Weigela* spp.)**

Climbers

Many climbers are excellent for bees, with the advantage that they can cover large areas of wall or fence that might otherwise offer little or no nectar or pollen. Honeysuckle (*Lonicera* spp.) is classically associated with bees, although only the garden bumblebee can get its tongue to the bottom of the extended flower tubes of most varieties. Native wild honeysuckle (*L. periclymenum*) has perhaps the prettiest flowers, and they are also particularly attractive to moths. Wisteria is glorious for a short few weeks, and bees keenly take advantage of it while they can. *Wisteria sinensis* is commonly grown on the front of houses and over pergolas, its long racemes of white or lilac flowers particularly appreciated by honey bees. Bumblebees seem to prefer *W. frutescens*, which has smaller clusters of purple flowers. Unlike their shrubby mop-headed cousins, climbing hydrangeas can attract huge numbers of honey bees and short-tongued bumblebees. *Hydrangea serratifolia* can reach great heights and is especially popular with bumblebees. Boston ivy (*Parthenocissus tricuspidata*) or Virginia creeper (*P. quinquefolia*) can turn the whole of the outside of your house into an enormous nectar bar, growing particularly well on otherwise challenging north-facing walls. In late summer, the inconspicuous flowers can be visited by so many industriously humming honey bees that it can sometimes sound as if the building is going to take off and fly away. A few months later the house will be lit up with autumnal yellows and reds.

Trees for bees

Trees can be particularly valuable as a source of pollen and nectar because of their size; a large tree in bloom can be like a meadow in the sky – perhaps the equivalent of several acres of flowers. Mature trees are especially important to honey bees, offering not only a significant source of forage but also a potential home. Decayed cavities in trees are a favoured nesting location for free-living honey bees, providing more security and warmth than almost anywhere else. True to its name, the tree bumblebee (*Bombus hypnorum*) also likes to nest in tree cavities. Other species of bumblebee often make a home at the base of large trees, where they nest in holes and burrows found among the roots. Old, dead trees are often filled with beetle holes, ideal nesting sites for species of small solitary bee species.

Wild trees

Native trees growing in woodlands and hedgerows are among the most important trees for bees because they are often present in significant numbers and can provide pollen, and perhaps a little nectar, in late winter and spring when little else is available. The first to flower each year is hazel (*Corylus avellana*)*. Its bright yellow lambs'-tail catkins, hanging on otherwise bare trees, can brighten even the dullest winter days. For beekeepers, there are few more heart-warming sights than bees bringing home the first loads of yellow-green hazel pollen – a sign that overwintered colonies are alive and starting to expand. After hazel comes willow, which can offer large amounts of pollen from catkins on male trees and also some nectar from the catkins on female trees. There are various species of willow, the most important for bees being the pussy or goat willow (*Salix caprea*)***, whose sherbert-yellow male catkins are highly conspicuous, usually from mid-March. Goat willow can be grown in the garden as a medium-sized tree, in a hedge or as a small, weeping tree that will produce only the pollen-bearing male catkins (*S. caprea* 'Kilmarnock'). Another good option for smaller gardens is the woolly willow (*Salix lanata*)**. Several species of solitary bees are willow specialists, including Clarke's mining bee (*Andrena clarkella*), the early colletes bee (*Colletes cunicularius*) and the small and large sallow mining bees (*Andrena praecox*, *A. apicata*).

There is a succession of spring-flowering wild trees offering mostly pollen, including alder, birch, elm, poplar, beech, oak and hornbeam. The first major sources of nectar are blackthorn (*Prunus spinosus*)*, damson (*P. domestica*)** and wild cherry, of which there are two species in the UK; the cherry (*Prunus avium*)***, and the bird cherry (*P. padus*)***.

Late April usually brings the flowering of hawthorn (*Crataegus* spp.)***. Hawthorn blossom can be a good place to spot various species of newly emerged solitary bees. Honey bees love hawthorn too, and in some years there can be enough nectar to give an early harvest of dark, strong-flavoured honey, although hawthorn

nectar flows are famously fickle. Hawthorn makes a characterful garden tree that is excellent for wildlife and is available in several cultivars and even as handsome multi-stem specimens. Also flowering in mid-spring can be field maple (*Acer campestre*)** and sycamore (*A. pseudoplatanus*)***, both of which have garden-friendly cultivars and can contribute significantly to a crop of spring honey. Wild crab apples and the various roadside apple trees (*Malus* spp.)** that result from discarded cores also flower in April, and are a reliable source of pollen and nectar.

Late spring brings horse chestnut (*Aesculus hippocastanum*)** into bloom. Its white, sometimes deep pink, candelabra flowers are visited for pollen and can provide lots of nectar, their pink centres turning yellow once they have been pollinated. Distinctive, brick-red horse-chestnut pollen is easy to spot on honey bees'

RIGHT Horse chestnut (top) and Goat willow (bottom).

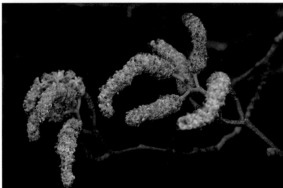

ABOVE Tree bumblebee (*Bombus hypnorum*) on lime tree flower.
LEFT Alder (top) and Sycamore (bottom).

legs as they enter the hive and when stored in the comb. Indian horse chestnut (*Aesculus indica*)** is a beautiful but large garden tree that flowers about a month later.

Undoubtedly, the most important UK tree for honey is the lime. Not to be confused with the tree that produces the citrus fruit, there are about 30 species of lime tree (*Tilia* spp.)***, the three most common in the UK being the broad- or large-leaved lime (*T. platyphyllos*), the small-leaved lime (*T. cordata*) and the common lime (*T. × europaea*). The sources of most lime honey are thought to be the broad-leaved lime, a large tree frequently planted in parkland, and the common lime, often planted in avenues and streets – making it important to urban beekeepers. Limes typically flower from mid-June for about a month and, in ideal conditions, can yield large amounts of exquisite honey with a delicate, minty-citrus taste. The unusual, pendant flowers hang from bracts and are worked by honey bees and short-tongued bumblebees, often early in the morning. Limes are large trees, but the Mongolian lime (*T. mongolica*)*** is smaller and suitable for modest gardens.

Garden trees

Many ornamental garden trees can be of great importance to bees. The autumn-flowering cherry (*Prunus × subhirtella* 'Autumnalis')** can flower sporadically throughout winter and into spring, when the more blousy pink and white flowers of most other ornamental cherries (*Prunus* spp.)*** take over. Apart from double-flowered forms, all flowering cherries are highly attractive to bees. Usually coming into flower at about the same time, snowy mespilus (*Amelanchier lamarckii*)** is a beautiful small tree that looks glorious when covered in ice-white flowers backed by copper-coloured young leaves. It's among the most popular small trees for a garden, and much loved by bees. For a vivid splash of spring colour, the Judas tree (*Cercis siliquastrum*)* has shocking-pink flowers, some of which grow directly from the trunk. On a warm day, they will be covered in bees.

Most summer-flowering trees are non-natives. Among the best, though rarely planted, is the seven son flower tree (*Heptacodium miconioides*)**. Once thought to be extinct, when in flower this Chinese tree can attract so many bees that it can sound as if a swarm has landed in it. Similarly popular with bees, the aptly named bee-bee tree (*Tetradium daniellii*)*** is highly recommended. Flowering in late summer and early autumn it provides nectar when most other sources have finished. It is a medium-sized tree, so is suitable for many gardens.

As well as nectar and pollen, trees are the main source of two other important resources for honey bees. Propolis is collected from many trees, sometimes from sap or resin exuded from the trunk and branches of trees such as firs, poplars and cherries, and sometimes from sticky flower buds, such as those of the horse chestnut. Trees are also the commonest source of honeydew, which is usually only gathered in large quantities in years when drought has reduced the nectar available from flowers, or when there are particularly large numbers of aphids.

It can be difficult to recommend bee trees for gardens because many mature trees would be unsuitable for the average, modest garden. Therefore, in the following list I have indicated where a tree is particularly suitable for smaller spaces. Many trees are available as cultivars that grow to different sizes, so check with a trusted tree nursery before you buy. Some trees are good providers

ABOVE Amelanchier.

only of pollen, being particularly important in spring. Some produce both pollen and nectar. These attributes are noted as follows: S = smaller tree, P = provides pollen, N = provides nectar.

Suggested early-flowering trees for bees:

- Alder (*Alnus glutinosa*) P**
- Blackthorn (*Prunus spinosa*) SPN**
- Blue wattle (*Acacia dealbata*) SPN***
- Beech (*Fagus sylvatica*) P*
- Cherry plum (*Prunus cerasifera*) SPN**
- Cherry – ornamental (*Prunus* spp.) SPN***
- Crab apple (*Malus sylvestris*) SPN***
- Elm (*Ulmus* spp. Note that *U. carpinifolia* and *U. glabra* 'Camperdownii' are smaller trees) P*
- Field maple (*Acer campestre*) PN***
- Hawthorn (*Crataegus* spp.) SPN***
- Hazel (*Corylus avellana*) SP***
- Judas tree (*Cercis siliquastrum*) SN**
- Winter-flowering cherry (*Prunus × subhirtella* 'Autumnalis') SPN***
- Oak (*Quercus* spp.) P*
- Snowy mespilus (*Amelanchier × lamarckii*) SPN***
- Snowdrop tree (*Halesia carolina*) PN**
- Sycamore (*Acer pseudoplatanus*) PN***
- Willow (*Salix* spp.) PN***
- Yew (*Taxus baccata*) P*

271

Recommended summer-flowering trees:

- Eucalyptus (*Eucalyptus* spp.) PN**
- False acacia (*Robinia pseudoacacia*) N**
- Golden rain tree (*Koelreuteria paniculata*) PN**
- Holly (*Ilex aquifolium*) SPN**
- Hop tree (*Ptelea trifoliata*) SN***
- Horse chestnut (*Aesculus* spp.) PN***
- Indian bean tree (*Catalpa bignonioides*) PN**
- Lime (*Tilia* spp.) PN***
- Mountain ash (*Sorbus* spp.) SPN**
- Seven son flower tree (*Heptacodium miconioides*) SPN***
- Siberian pea tree (*Caragana arborescens*) SPN**
- Silk tree (*Albizia julibrissin*) SPN**
- Swedish whitebeam (*Sorbus intermedia*) PN**
- Sweet chestnut (*Castanea sativa*) PN***
- Tree of heaven (*Ailanthus altissima*) PN***

Suggested late-flowering trees:

- Japanese angelica tree (*Aralia elata*) SPN**
- Japanese pagoda tree (*Sophora japonica*) PN**
- Strawberry tree (*Arbutus unedo*) SPN**
- Bee-bee tree (*Tetradium daniellii*) PN***

LEFT TO RIGHT FROM TOP:

THIS PAGE Judas tree; Crab apple; Blue wattle, Eucalyptus, Bee-bee tree.

OPPOSITE PAGE Laburnum, Swedish whitebeam; Mountain ash, Sweet chestnut; Seven son flower tree, False acacia.

Garden and farm crops

A very practical reason for protecting pollinators is the role they play in the production of much of our food. But pollination can sometimes seem like a somewhat abstract concept – something that happens in far-off places that somehow produces the kiwi fruit and avocados on our supermarket shelves. However, if you grow your own fruit and vegetables, you can witness pollination as it happens in your garden, and benefit from the results.

The importance of having sufficient bees of the right kind to pollinate a crop is recognised by farmers around the world, who often pay to have bees brought to their fields at flowering time. The best-known example is the annual almond pollination event in northern California. Every February, around two million hives of honey bees are shipped sometimes thousands of miles, to spend a month among the almond groves. The result is the pollination of 1.7 million acres of trees and a harvest of about one million tonnes of almonds – 80 per cent of world production. When they have finished on the almonds, the bees are taken on a merry-go-round trip throughout the USA, visiting Washington to pollinate apples, Kansas for sunflowers, Texas for melons, Georgia for blueberries and Wisconsin for cherries – with a few other stops along the way. It's an effective way of pollinating crops on an industrial scale, but the consequences for the bees are sobering; a huge proportion (often over 60 per cent) of colonies die over winter, mostly as a result of stress-related issues. UK commercial pollination is on a much more modest scale, with a few beekeepers being paid to take small numbers of hives to crops such as apples, strawberries and occasionally field beans.

If you grow fruit and vegetables, you might never consider how they are pollinated or even *if* they are pollinated – it's something that usually just happens.

However, you will probably notice the times when crops are insufficiently pollinated. In some years, harvests of fruit from early-flowering trees, such as plums or apricots, will be very poor. This can be because spring was cold or wet, keeping bees away from the blossom. You might also sometimes notice wonky apples, lumpy strawberries or pears that aren't very pear-shaped. Sometimes squash and courgette fruits start to grow and then wilt and shrivel away. Again, these are signs of inadequate pollination.

The majority of the productive plants that you are likely to grow depend on pollination to one extent or another. In fact, of the 170 agricultural crops on which pollinator research has been published, 84 per cent need pollinators. In some cases, flowers need to be pollinated for a harvestable, edible fruit to form. With others, a pollinator must have visited a flower for it to produce the seed that you then plant and grow.

By attracting a wide range of bees to your garden you will improve the pollination and productivity of your edible crops. Bumblebees are especially useful for pollinating early-flowering fruit like plums, cherries and gooseberries. Their built-in fur coat means that they can work at lower temperatures, and visit flowers on cold spring days when few other bees are flying. Bumblebees and some solitary bees are also capable of pollinating flowers that require buzz pollination, or sonication. The flowers of plants including tomatoes, aubergines, potatoes and blueberries have tubular anthers with an opening at one end, their pollen firmly attached to the inner walls of the tube. Wind may shake loose some of the pollen, and the clumsy efforts of some bees trying to gain access might do the same, effecting a degree of pollination. But much more efficient pollination is achieved by bees that know how to properly dislodge the pollen by wrapping their legs around the anthers and vibrating their flight muscles. This dislodges the pollen and sends it cascading out of the tube and onto the bee.

LEFT Honey bee on a strawberry flower. Every pip on the outside of a strawberry is the result of pollination.

ABOVE Bees can benefit from pollination too, feasting on the sugary juice of fallen fruits.

RIGHT Imported nests of bumblebees in a Kent cherry orchard.

The ability of bumblebees to fly at low temperatures and to buzz pollinate like this means that they have great economic value, and so in some areas they are bred commercially for sale to growers. Colonies of buff-tailed bumblebees (*Bombus terrestris*) in cardboard nest boxes are placed in orchards or inside greenhouses where crops need to be pollinated. When they have outlived their usefulness, the nests and bees are destroyed. It is becoming apparent that in some situations bumblebees are less effective at buzz pollinating than they should be; research shows that neonicotinoid insecticides inhibit their ability to perform this important task.

Some solitary bees are particularly efficient pollinators because of their somewhat inefficient way of collecting and carrying pollen. Bumblebees and honey bees gather pollen on their hairy coat and then then dampen it with a little nectar, compacting it to form a neat blob that can be carried in their pollen baskets. Solitary bees don't have pollen baskets; instead, they rummage among the anthers of a flower to pick up as much pollen as they can, letting it lodge somewhat messily among their stiff hairs. They then fly to another flower and repeat the process, and the dry grains of pollen are easily dislodged and transferred. It has been estimated that some solitary bees can be up to 120 times more efficient at transferring pollen than a honey bee,

so fewer are needed for effective pollination. Solitary bees such as the red mason bee (*Osmia bicornis*) are particularly useful for pollinating crops like apples, and are sometimes cultivated and released in orchards for this purpose. In the USA, the orchard mason bee (*O. lignaria*) is cultivated on a large scale for fruit pollination.

Where honey bees can't be beaten is in their sheer numbers and convenience. A single colony will comprise tens of thousands of hard-working bees, all living in a convenient box that can be placed wherever pollination is needed and taken away again when the job is done. As a small-scale beekeeper, you probably won't move your bees in this way, but if you keep honey bees at home then – along with the wild species of bees you attract to your garden – they will ensure thorough pollination and the production of good-quality fruit and vegetables.

Bees and other pollinators are particularly important where cross-pollination is necessary. Apples are an intriguing example of this; most varieties are not self-fertile, so the flower on each tree needs to be fertilised with pollen from a tree of another variety planted nearby – ideally, within 20 metres. The snag is that different apple varieties flower at different times. To ensure that trees are compatible they are divided into seven groups according to flowering time, meaning you can select compatible varieties likely to flower simultaneously. Some varieties are even triploid, needing a three-way relationship. If you live in a highly populated area, the chances are that nearby gardens will have trees whose pollen can fertilise your own trees, and vice versa. But if you live somewhere remote, you might have to play Cupid to make sure your apple trees are well matched. Strawberries also have demanding pollination requirements. They are self-fertile so don't need pollen from another flower, although cross-pollination will boost the flavour and sugar content of the fruits. But wherever it comes from, they do need to have lots of pollen delivered to their stigmas. This can be done with pollen wafting in the air but is much more effective if helped by pollinators. A strawberry flower has 50–200 stigmas, and it can take 6–15 bee visits to make sure pollen is delivered to them all. The result of good pollination is a fat,

well-rounded berry covered on the outside with many tiny seeds – each one the result of being pollinated and fertilised. If insufficiently pollinated, fewer seeds will set and the berry will be small, misshapen and contain less sugar. Whether plants are self-fertile or need cross-pollination, pollinators facilitate the effective transfer of pollen, producing better quality, heavier crops.

Fruit crops that benefit from pollinator visits:

- Apples
- Apricots
- Blackberries
- Blueberries
- Cherries
- Currants (red and white)
- Gooseberries
- Loganberries
- Melons
- Medlars
- Peaches
- Pears
- Plums, damsons and gages
- Quinces
- Raspberries
- Strawberries

BELOW The red mason bee is an important pollinator of apples.

BELOW Gooseberry flower being worked by a honey bee.

The vegetable plot

Once, flowers were grown in one part of the garden and vegetables in another, and the separation was absolute. Today, however, the boundaries are often more blurred. This makes good sense, as not only are many vegetable plants extremely beautiful – think of the colourful stems and glossy leaves of rainbow chard – but having ornamental flowers immediately next to vegetables can attract more pollinators, ensure improved pollination and produce a better crop. Another advantage is that flowers can attract various useful predators such as lacewings and ladybirds which, along with their larvae, can eat their way through vast quantities of aphids and other vegetable plot undesirables.

Flowers grown on a vegetable plot are best planted so that they flourish once the vegetable plants have established and are growing well. Low-growing flowers should be used so that they don't shade your vegetable plants. Pot marigolds (*Calendula officinalis*)** are excellent for attracting pollinators, while nasturtiums (*Tropaeolum majus*)** are loved by long-tongued garden bumblebees (*B. hortorum*) and help to draw aphids and blackfly away from your beans. Crimson clover (*Trifolium incarnatum*)** or the tiny-flowered black medick (*Medicago lupulina*)* will both lure pollinators, fix nitrogen in the soil and suppress weeds. Even the tiny flowers of chickweed (*Stellaria media*)*, commonly considered an annoying weed, can attract bees to your vegetables.

ABOVE Bumblebee and honey bee visiting a courgette flower.

Vegetable crops that benefit from pollinator visits:

- Aubergines
- Beans – most kinds
- Globe artichokes
- Jerusalem artichokes
- Okra
- Squashes, marrows and courgettes
- Sweet peppers
- Tomatoes

Vegetable flowers and seeds

Many vegetable plants don't require pollination to produce a crop, because we eat the stems, leaves or roots of the plants before they themselves flower. When vegetables do begin to flower it is usually considered the end of their productive life, but actually they still have a lot going for them. Rather than digging them out, let them flower, as most are rich in pollen and nectar and are highly attractive to bees and other pollinators. Some are annuals and will flower in the year of planting, such as lettuce. Others are biennial and will flower the flowering year, like many of the brassicas and root vegetables. Leave biennials in the ground over winter, let them flower in spring and then remove them in time for the new season's plantings. You don't have to let a whole crop flower – just one or two plants on the edge of a plot will help.

It's not only pollinators that can benefit from allowing vegetables to flower. Once pollinated, many will produce seeds that you can harvest and plant. Fresh seed usually germinates much better than seed bought in packets, and will save you considerable expense. Leeks, parsnips, beetroot, lettuce, broccoli and kale are some of the vegetable plants whose seeds you can easily harvest and grow. Be aware that seeds from F1 hybrid varieties will not come true and produce plants exactly like their parents. Furthermore, if you grow more than one variety of the same crop they might cross-pollinate and produce offspring of an unknown quality.

All of the alliums produce beautiful flowers that are tremendously attractive to bees. There is a tendency to pull out leeks and onions as soon as they start to flower, but leave them for the bees and see how they flock to them. For garlic to be useful to bees, plant the hardneck varieties. These grow a flower stem called a scape, which can be harvested and eaten or left to produce a flower that is loved by bees. The large amount of energy used producing the flower means that the garlic bulb will be smaller, so let a few plants flower and then harvest a selection of different-sized bulbs.

Parsnips left in the ground will flower, but flowers can also be grown from the parsnip tops left over after preparing your Sunday roast. Sprout the tops in a saucer of water and plant them out when they have roots.

They will grow feathery leaves and produce yellow, umbelliferous flowers that are especially popular with small solitary bees.

Plants that produce useful flowers after going over:

- Alliums: leeks, onions and garlic
- Asparagus
- Beetroot
- Cabbages, kale and brussels sprouts
- Carrots
- Celery
- Chicory
- Globe artichoke
- Lettuce
- Mustard
- Parsnips
- Radishes
- Rapini
- Rocket
- Turnips

LEFT AND BELOW Two common garden vegetables left to flower for the bees: onions and asparagus.

ABOVE Honey bee on chickweed.

Green manures

Green manures can be sown on bare soil as soon as a crop has been removed. They help to prevent rain from causing damage to the soil structure and leaching out nutrients. Once grown, they are dug into the soil to improve its structure and fertility. Various plants can be used as green manure, one of the most commonly used being phacelia (*Phacelia tanacetifolia*), which is also among the best of all bee plants if allowed to flower. Various green manure seed mixes are available, with the Tubingen Mix being specifically designed for bees. It contains black cumin, borage, buckwheat, calendula, coriander, cornflower, dill, mustard, oil radish, phacelia and wild mallow. Sow the seed at any time between March and September, as soon as any vegetable crop has been removed. The plants should remain in the ground long enough for them to flower and benefit bees and other pollinators, before being dug into the soil three to four weeks before a new vegetable crop is planted. If you let green manure crops flower for bees, it's a good idea to remove developing seedheads so that they don't produce thousands of seedlings that crowd out your crops.

Wonderful comfrey

Comfrey is considered a herb, but it shouldn't be eaten. It has many medicinal uses and is also sometimes called knitbone because of its healing properties (though it should only be used externally). It is a statuesque plant with beautiful flowers, but is rarely grown in an ornamental border because it is also a terrible thug. No one ever seems to know quite where to plant it, so often it is assigned to a difficult corner, or next to the compost heap. Comfrey is tremendously useful as a fertiliser; fill

ABOVE Female hairy-footed flower bee approaching a comfrey flower.

a bucket with leaves, add water and stand well back. The stench is vile, but after a week or two the resulting liquid, diluted about 1:4, is a well-balanced plant feed. Perhaps comfrey's greatest asset, though, is its appeal to bees. Long-tongued bees adore the deep, nectar-rich flowers that bloom from mid-spring into summer. Short-tongued bees will nibble holes in the base of the flowers and steal nectar, with honey bees joining the queue.

Common comfrey (*Symphytum officinale*)*** can form enormous clumps and self-seed everywhere. For something better behaved, opt for the sterile hybrid *Symphytum × uplandicum* 'Bocking 14'. Cut back to the ground in early summer and use the whole plant for fertiliser or compost, even just heaping it around trees and shrubs. The stump will regrow and produce late-summer flowers, ideal for queen bumblebees preparing to hibernate. For a more restrained plant, perfect for growing along the shady edge of a fence or hedge, try dwarf comfrey (*S. grandiflorum*)*** or Iberian comfrey (*S. ibericum*)***, which grow only to about 50cm.

Herbs for bees

Two thousand years ago, Virgil wrote about honey bees and hives, saying: 'Let green rosemary and wild thyme with far-flung fragrance, and a wealth of strongly-scented savory, flower around them.' If you holiday in the Mediterranean today, you will be able to sample rich, aromatic honeys made from the nectar of herbs such as thyme, rosemary, lavender and sage – few honeys are so redolent of the landscape that produced them. Bees flock to flowering herbs almost more than to any other kind of forage, and it is worth planting as many as you can – though unless you do so by the acre, you are unlikely ever to harvest honey of the sort Virgil might have enjoyed.

The definition of a herb is somewhat hazy. Strictly speaking, it is any herbaceous plant – one that dies down to the ground over winter. But generally a distinction is made between herbaceous plants grown only for ornament, and herbs that are in some way edible or medicinal. Here I have divided them simply into culinary herbs, for those commonly used in the kitchen, and other herbs, for all those generally considered to have beneficial properties or which are difficult to categorise otherwise.

Culinary herbs

With their heavy scent and sugar-rich nectar, the flowers of many culinary herbs are hugely attractive to bees. Growing them in and among fruits and vegetables will help lure more bees and improve pollination and crop production. Not all culinary herbs are Mediterranean, but many of the best-known ones are. They are happiest in a sunny position with very well drained soil, meaning they are likely to become increasingly important bee plants as our summers get hotter and drier. Many herbs are members of the Lamiaceae (mint family), a large group of flowering plants almost all of which are important for bees – including, rather surprisingly, the teak tree. Other members include the lavenders, salvias, thymes and deadnettles, all of which have petals fused into tubular flowers with an upper and lower lip.

Rosemary (*Salvia rosmarinus*)*** is the source of the renowned French Narbonne honey. Its flowers are useful at times when little else is available, often blooming from late winter until early summer, again in early autumn and sometimes also for periods in between. Honey bees, bumblebees and some solitary bees use it. It can be grown as an evergreen shrub in the border, but also works well in pots. Prostrate varieties will spill down steps and over walls, turning otherwise unused spaces into foraging hotspots. It has lovely lavender-blue flowers, but if pink is your thing, try *S. rosmarinus* 'Rosea'.

ABOVE A honey bee with distinctive purple-grey rosemary pollen on her back legs.

Like rosemary, sage is a member of the salvia family and equally beloved by bees. Most people grow common sage (*Salvia officinalis*)** which has deep flowers enjoyed by long-tongued bees. *S. lavandulifolia* is a smaller plant, ideal for containers, and has shallower flowers that are more useful to shorter-tongued bees.

Thyme (*Thymus* spp.)*** is another extremely useful herb for bees. There are hundreds of varieties and, if you pick the right combination, it can be in flower from May until October. It comes in three forms: upright thymes are bushy with easily picked stems, ideal for cooking; creeping thymes are very low-growing and can make a carpet of flowers across paths or lawns; and mounding thymes grow in rounded forms – different cultivars grown together will merge into undulating hummocks of pink, purple and blue flowers. Thymol, an essential oil produced from thyme, is a key ingredient in some varroa treatments used in honey bee colonies and it is thought that honey bees might self-treat by visiting thyme flowers, which they do in great numbers. Thyme honey has a slightly medicinal taste that suggests it is doing you some good.

Oreganos and marjorams are of the same genus but are different species, causing much confusion. They are superficially very similar and bees visit both with equal vigour when in flower. Wild marjoram (*Origanum vulgare*)*** and wild thyme (*Thymus polytrichus*)*** grow extensively in areas of the UK with chalk soil and can contribute to a honey crop.

Mints (*Mentha* spp.)*** when in flower are hugely attractive for bees, particularly honey bees which after visiting can produce honey with a lightly minty tang but which loses its flavour with age. You are unlikely to experience this honey from your garden bees as, if you are wise, your mint will be restrained in pots to keep it from running rampant. Water mints growing near rivers or in marshy areas usually flower from August so are a useful late-season source of nectar. The hairy stems of some mints are sometimes collected by wool carder bees (*Anthidium manicatum*) for their nests.

Lemon balm (*Melissa officinalis*)** is closely linked to bees, not least because its genus name, *Melissa*, comes from the Greek for honey bee. Honey bees do visit it when it is in bloom, but the deep flowers better suit longer-tongued bumblebees. However, the smell of lemon balm is highly attractive to honey bees, and crushed leaves rubbed inside an empty hive or skep are said to encourage swarms to take up residence. Lemon balm essential oil is more powerful, and a few drops sprinkled from a bottle are better at enticing bees than the fresh leaves. Lemongrass oil is best of all. Some beekeepers grow lemon balm near their

ABOVE Wild marjoram and a common carder bumblebee in a wildflower meadow.

hives and rub their hands on it before inspections, believing the smell calms bees and reduces stings. If you do get stung, summer savory (*Satureja hortensis*)** could be useful as it has traditionally been rubbed on skin as a sting relief. Both summer savory (an annual) and winter savory (*S. montana* – a perennial)*** are very heavily worked when in flower by honey bees and short-tongued bumblebees. They make a useful, shorter alternative to lavender as an edging plant.

Hippocrates recommended hyssop (*Hyssopus officinalis*)** for chest complaints, but you are probably better off simply enjoying the sweet anise flavour of the beautiful purple flowers in a colourful salad. Bees, on the other hand, gorge themselves on the nectar while picking up bright orange pollen from the spiky anthers.

Coriander (*Coriandrum sativum*)** is mostly associated with Asian cuisine. Sow successively for a supply of leaves throughout the summer. It bolts easily, giving bees a chance to use the tiny, pretty white flowers, and then produces seeds that you can gather for the kitchen. Coriander honey, harvested where crops are grown on a commercial scale, has a lemony-citrus flavour.

Like all alliums, chive flowers are irresistible to bees. Treat chives (*Allium schoenoprasum*)** as an edging plant around your vegetable beds – they are said to improve soil and protect other plants from various pests. They

ABOVE Thymes are loved by bees and might play a part in maintaining colony health.

certainly attract plenty of useful pollinating bees. After flowering, cut them back for another flush. Garlic chives (*Allium tuberosum*)*** are astonishingly attractive to bees and fortunately their smell and taste don't seem to make it into the honey.

Probably more of a spice than a herb, and with a scientific name that makes it sound like a laboratory-created pharmaceutical drug, Sichuan pepper (*Zanthoxylum simulans*), is an unusual and fun addition to the garden. Growing as a bushy, spiny, medium-sized shrub, it produces sprays of tiny, yellow-green flowers in early summer. Only flowering for a few days, it could hardly be said to be an important forage plant, but while it is in flower it seems to entice all the bees in the district, making the branches vibrate as they work the flowers. If you grew enough bushes perhaps you would get spicy honey, but otherwise you can gather and dry the seeds and use them in the kitchen.

Other culinary herbs to grow for bees:

- Basil (*Ocimum basilicum*)**
- Bay (*Laurus nobilis*)*
- Black cumin (*Nigella sativa*)***
- Dill (*Anethum graveolens*)*
- Fennel (*Foeniculum vulgare*)*
- Lovage (*Levisticum officinale*)**

ABOVE Like all alliums, chive flowers are adored by bees.

LEFT TO RIGHT FROM TOP Winter savory, Oregano; Hyssop, Thyme; Sage, Coriander; Mint, Rosemary.

Other herbs

Lavender (*Lavandula* spp.)*** is the quintessential bee plant, and it is impossible to picture a haze of purple-blue lavender spires without clouds of bees buzzing merrily around them. Lavender is mostly grown for its richly scented perfume and is generally not much thought of as a culinary herb, although it can be used in the kitchen in all sorts of ways, including in a luscious lavender and honey ice cream. Lavender offers nectar freely from the many tubular florets that form each flower head and it is visited by bees of all kinds, although some show a preference for certain cultivars more than others – probably because their tongue lengths better match the depth of their flowers. The so-called French lavender (*L. stoechas*), with bunny-ear bracts, is not favoured nearly as much as English lavender (*L. angustifolia*) of which 'Hidcote' and 'Munstead' are among the most popular with bees and gardeners alike. Trials by the University of Sussex found a Dutch lavender, *L. × intermedia* 'Gros Bleu', to be the best of all in terms of visits from bees, with the added advantage that it flowers for longer than most other cultivars. *L. × intermedia* 'Edelweiss' has been found to be the most attractive white-flowered lavender. Crops of the beautifully perfumed honey are rare in the UK, even where lavender is farmed, because the flowers, harvested for their oil, are often cut in their prime.

Although superficially similar to lavender, nepeta or catmint (*Nepeta* spp.)***, is, if anything, even better as a bee plant. Its stems hold flowers from top to bottom and its rich nectar is devoured by more species of bees than lavender. Blooming slightly later than lavender but for much longer, it is equally as good in hot, dry conditions. Lavender probably only wins in the popularity stakes (with humans, that is) because of its heavenly scent and more dignified deportment – catmint can flop a bit, especially after felines have had a roll in it. There are lots of cultivars, from big sprawlers like 'Six Hills Giant' to the neat and neatly named *Nepeta × faassenii* 'Purrsian Blue'.

Borage (*Borago officinalis*)*** is yet another superb bee plant, said to produce more nectar than any other garden flower. The clear blue star-shaped flowers, good in a salad or a glass of Pimms, form an umbrella that protects nectar from both rain and sun, ensuring it is

ABOVE Beautiful, fragrant, bee-friendly lavender. This is *Lavandula angustifolia* 'Hidcote'.

always available. It is a free-flowing nectar bar; almost as soon as nectar is removed by bees, more arrives to replace it. Borage is a fast-growing annual and ideal for quickly filling any gaps in the border; put in a seedling or two and they can be flowering in a matter of weeks. They spread themselves around a bit, but are easily pulled up and not much of a nuisance. Borage is grown commercially to produce starflower oil and, if you are lucky to have fields of it nearby, you might get lots of very light-coloured honey with a slightly herbal taste.

Delightfully understated with blue-grey foliage and dusky-pink flowers, germanders (*Teucrium* spp.)*** don't go unnoticed by bees, which work them fastidiously throughout summer. Honey bees and short-tongued bumblebees love them and a patch of wall germander (*T. chamaedrys*) that I watch regularly is used by green-eyed flower bees (*Anthophora bimaculata*) and is a favoured mating site for wool carder bees (*Anthidium manicatum*). Hedge germander (*T. lucidrys*) is equally popular while wood sage (*T. scorodonia*), with unusual pale yellow flowers, is worth growing too. Caucasian germander (*T. × hircanicum*) is the show-off of the family, with tall spires of reddish-purple flowers that always attract the attention of lots of bees.

Flowering from about late May until the first frosts and with dramatic flower spikes in white, burnt orange, pink, purple and blue, agastaches (*Agastache* spp.)*** are tremendously useful garden plants and are loved by bees. The deep-flowered orange species originate in Mexico and the southern states of the USA, where they are pollinated by hummingbirds, but in the UK they are used by long-tongued bumblebees such as the garden bumblebee (*Bombus hortorum*). The blue and purple varieties have shallower flowers and are enjoyed by other bumblebees and honey bees. On warm days, the leaves release their minty, aniseed aroma and perfume the garden.

For something statuesque but short-lived, plant one of the angelicas (*Angelica* spp.)***. They can grow to about 2m (6ft), with huge, domed heads that attract masses of insects, from large bumblebees to tiny solitary bees, as well as wasps, and hoverflies. The caterpillars of several species of moths and butterflies feed on the leaves. Grow angelica from seed and it will flower after two or three years and then die. *A. gigas* is an unusual purple colour.

Other flowering herbs loved by bees:

- Arnica (*Arnica chamissonis*)**
- Calamint (*Calamintha nepeta*)***
- Chicory (*Cichorium intybus*)**
- Coneflower (*Echinacea* spp.)***
- Evening primrose (*Oenothera biennis*)**
- Mallow (*Malva* spp.)**
- Motherwort (*Leonurus cardiaca*)**
- Pot marigold (*Calendula officinalis*)**
- Skullcaps (*Scutellaria* spp.)**
- St John's wort (*Hypericum* spp.)*
- Tansy (*Tanacetum vulgare*)**
- White horehound (*Marrubium vulgare*)***
- Wild bergamot (*Monarda fistulosa*)**

BELOW Female red mason bee approaching nepeta.
OPPOSITE PAGE, LEFT TO RIGHT FROM TOP Hedge germander, Bastard balm; Tansy, Agastache, Evening primrose; Lavender, Coneflower, Wood sage; Angelica, Marsh mallow, Borage.

The agricultural landscape

Some 70 per cent of land in the UK is classed as agricultural and so is potentially of immense importance to wildlife of all kinds. If you keep bees in the countryside, knowing what's being grown on nearby farmland is important because it can affect how your colonies might grow and develop, and the way in which you manage them.

Agricultural crops can be a rich source of nectar and pollen that can boost colony growth and produce good crops of honey. But there is always a lurking fear that any agrochemicals used to treat those crops could cause harm to your honey bees or to local wild bees. Fortunately, large-scale bee deaths caused directly by pesticides are rare in the UK, mainly because their use is strictly regulated. A more likely scenario is that bees visiting treated crops or nearby wildflowers may suffer from hard-to-identify, sublethal conditions that can affect their overall general health and lead to gradual population declines, particularly in the case

of unmanaged wild bee species. Broadly speaking, however, it is perfectly possible to keep honey bees in rural areas without having to be unduly worried about potential devastation caused by agrochemicals. Usually, the greatest concern for countryside beekeepers is the overall lack of wildflowers to be found in agricultural landscapes. The most important and consistently reliable flower in rural areas grows in the hedgerows, overgrown field corners and anywhere else not too brutalised by the hedge trimmer. Brambles, or blackberry (*Rubus* spp.)*** flowers from early- until late summer. Their deep roots mean that even in dry weather they can produce nectar in abundance. They are a tremendously varied plant with more than 1,000 micro-species, meaning the plants are slightly different nearly everywhere you go. Many species of bees visit the flowers but, for honey bees, brambles are invaluable; more UK honey is produced from bramble nectar than from probably any other source.

Cereals such as wheat and barley offer nothing to bees but are usually grown on a rotational system, being replaced with a break crop every four or five years to

ABOVE Bramble is among the most important wild flowers for bees. This is a male woodcarving leafcutter bee (*Megachile ligniseca*).

ABOVE Beehives next to flowering field beans – a crop of delicious honey is possible if the weather is kind.

RIGHT Honey bee on oilseed rape.

help improve soil fertility and control the build-up of pests and diseases. Break crops usually produce a harvest, examples being oilseed rape or field beans. Sometimes a cover crop is grown – short-term plantings between annual cereal crops that usually don't produce a harvest but help improve soil condition and control weeds. Both break crops and cover crops can flower, so wherever cereals are grown there is the chance of an occasional field of flowers that might be useful to bees and other wildlife.

One problem with flowering agricultural crops is that they tend to result in either a feast or a famine, producing a great abundance of pollen and nectar during a few weeks of flowering, after which there may be nothing useful on that land for perhaps several years. In intensively farmed areas, honey bees are best moved close to crops like oilseed rape to take advantage of the short, intense flowering period and then away again once flowering is over, to allow wild bee species to benefit from what few wildflowers might be found in field margins and hedgerows.

The type of crops grown on farms varies according to local climate and soil conditions, but is also influenced by continually changing legislation, agricultural practices and market forces. Useful crops for bees might be grown for many years then quite suddenly be replaced by something more profitable. If you keep bees in the countryside, you may find yourself changing your management practices every so often to keep pace with agricultural trends.

ABOVE Borage can produce masses of light-coloured honey. Beekeepers are lucky if it is grown nearby.

BELOW Interesting plant mixes are sometimes grown as green manure or for game cover, often also benefitting bees. This mix includes buckwheat, mustard and radish.

Crops for bees

In most areas, oilseed rape or OSR (*Brassica napus*)***
is the most significant agricultural crop for nectar and
pollen. It lights up the countryside in springtime with
acres of fluorescent-yellow flowers, later producing
pods containing tiny black seeds that are crushed
to make vegetable oil and animal feed. Changes in
agricultural practice and the European Union-wide
ban on neonicotinoid insecticides have affected the
way OSR is grown in the UK, in some areas decreasing
how much is planted and altering the time at which it
flowers, although this situation is likely to be variable.
Typically, it is sown in autumn to flower in early to mid-
spring. Like all brassicas in flower, OSR is very attractive
to bees of all kinds and if it is nearby, and the weather
is favourable, your bees might produce a very large crop
of mild-tasting honey. The downside for beekeepers is
that this honey can crystallise very quickly, making it
difficult to extract (see page 117). Colonies can build up
very quickly on OSR, so be vigilant for signs of swarm
preparations at about this time.

Field beans (*Vicia faba*)*** are essentially broad beans
grown on a large scale. They are used for animal feed
but also exported to North Africa and the Middle East
for human consumption. Long-tongued bumblebees can
reach into the deep flowers to access nectar, while short-
tongued bumblebees and honey bees will rob honey
from holes nibbled in the base of flowers. Nectar is also
produced from extrafloral nectaries located on small
leaf-like structures called stipules, that grow on the stem.
Beans usually flower in late spring/early summer, filling
the air with a sweet perfume. If the soil is moist and the
weather is warm, lots of pleasant-tasting honey can be
produced. One downside is that honey bees have a habit
of storing the distinctive grey pollen in super frames,
directly next to the honey. This can mean the comb
remains clogged with pollen after the honey has been
extracted, the clogged cells having to be cut out or the
whole comb replaced. Farmers are usually happy – and
occasionally even pay – to have hives placed in or near
their bean fields, because good pollination can increase
yields significantly. Other agricultural crops of benefit
to bees tend to be grown more sporadically according to
local conditions and farming practices. In some areas,
fields of bright blue borage (*Borago officinalis*)*** are
grown for the production of starflower oil, used in health

ABOVE UK beekeepers might be able to harvest more sunflower
honey in the future.

supplements. Buckwheat (*Fagopyrum esculentum*)** is
increasingly grown as a cover crop, its white flowers
attracting many species of bees and sometimes giving a
harvest of dark honey. Phacelia (*Phacelia tanacetifolia*)***,
grown for the same reasons, is one of the best nectar
plants for honey bees. Finding its deep purple pollen
stored in comb is a giveaway that phacelia is growing
nearby. Sunflowers (*Helianthus annuus*)** are sometimes
grown in the UK and provide pollen and, in the right
conditions, lots of nectar. Sunflower honey is only
reliably produced in warm countries, but could become
more common in the UK with climate change.

Crops grown for forage as an alternative to grass
include red clover (*Trifolium pratense*)**, which is
particularly good for long-tongued bumblebees,
lucerne or alfalfa (*Medicago sativa*)*, which can attract
honey bees, bumblebees and solitary bees, and sainfoin
(*Onobrychis viciifolia*)***, which when grown in quantity
can be used by honey bees to produce a beautiful, deep
yellow honey. Other crops which might be grown locally
and will benefit bees include orchard fruit and soft fruit
(see page 277).

Some flowering crops that might appear to promise
forage for bees can be of disappointingly little use.
These include linseed or flax (*Linum usitatissimum*),
maize (*Zea mays*) and field peas (*Pisum sativum*).

The best garden plants for bees

A seasonal selection of the 100 best garden plants for bees

When planning a garden for bees it can be difficult to know which plants to choose and what bees or other pollinators might visit them. The following is a guide to the 100 best ornamental garden plants to attract a wide range of species.

Plants that flower in the summer are numerous and diverse, making it relatively easy to curate an attractive bee-abundant garden. The best method is to plant relatively few species but to do so in volume. Drifts of flowers will attract more bees and help them to forage efficiently. Quantity is important at this time as this is when bee populations reach their peak, with many individuals foraging for pollen and nectar to provision their nests and raise their young.

Late winter and early spring are crucial seasons for bee-friendly planting. At this time there aren't many bees around – perhaps a few queen bumblebees, early emerging solitary bees and honey bees. Because there are few bees, quantity is less important than variety and the availability of something suitable in flower whenever it is warm enough for bees to fly. Flowers at this time can be life savers, sustaining the bees that will later flourish and nest in your summer garden.

The plants suggested here are suitable for a wide range of bees with differing requirements. If you are a beekeeper wishing to maximise your honey crop, the bad news is that the average-sized garden bursting with flowers will make little difference to what a colony can produce – it takes visits to tens of thousands of flowers to create one jar of honey. For honey production alone you would be best to invest in a few acres of land and plant nothing but phacelia and borage surrounded by hedgerows of spring-flowering trees. Allowing ivy to grow in or around your garden will make a positive difference in autumn, however.

The following list contains some of the most useful plants grouped by flowering season. Start by selecting the important winter- and spring-flowering plants and build on this with your choices for summer. If you want to plant trees, do so as soon as you can; they can take some years to grow to a productive size. Before shrubs and perennials become established, fill any gaps with annuals and biennials (see page 231) to maximise available blossoms.

Climate change makes it increasingly difficult to say when exactly plants will come into flower and for how long; some plants can now be seen flowering well outside their traditionally allotted time. The plants here are listed in the season that their flowers are of most importance for bees. For example, heleniums are listed as autumn plants because although they typically come into flower in mid-summer they might bloom until the first frosts, making them valuable autumn plants for bees. Similarly, anchusa might flower throughout summer but is most useful to bees in late spring.

Following each species there is information about which bees a plant is most likely to benefit. Up to three coloured stars are awarded according to likely usefulness to each species. These are:

* ** ***   Honey bees

* ** ***   Solitary bees

* ** ***   Long-tongued bumblebees

* ** ***   Short-tongued bumblebees

Refer to the section on bumblebee species (see page 164) to find their tongue length.

Plant varieties found to be particularly favoured by bees are noted in squared brackets.

LEFT Honey bee approaching snake's head fritillary, proboscis at the ready.

* ** ***	Honey bees		
* ** ***	Solitary bees		
* ** ***	Long-tongued bumblebees		
* ** ***	Short-tongued bumblebees		
	T = Tree		

Spring (*March, April, May*)

	Honey bees	Solitary bees	Long-tongued bumblebees	Short-tongued bumblebees
Alkanet (*Anchusa* spp.)	**	*	**	**
Alliums – ornamental (*Allium* spp.)	***	*	**	**
Apple & crab apple (*Malus* spp.) **T**	***	***	**	**
Bastard senna (*Coronilla valentina*)	**	**	***	**
Blue wattle/mimosa (*Acacia dealbata*) **T**	**	**	**	
Californian lilac (*Ceanothus* spp.)	**	*	**	**
Camassia (*Camassia* spp.)	*	*		*
Cherry (*Prunus* spp.) **T**	***	***	**	**
Elaeagnus (*Elaeagnus umbellata*)	*	*	*	*
Flowering currant (*Ribes* spp.)	**	**	*	**
Forget-me-not (*Myosotis* spp.)	**	**	**	**
Fritillaries (*Fritillaria* spp.) [*F. imperialis, F. meleagris*]	*	*	*	*
Grape hyacinth (*Muscari* spp.)	**	*	*	*
Hawthorn (*Crataegus* spp.) **T** [*Crataegus × lavallei* 'Carrierei']	***	***	*	*
Honeywort (*Cerinthe major* 'Purpurascens')	**	**	***	**
Leopard's bane (*Doronicum* spp.)	**	*	*	**
Lungwort (*Pulmonaria* spp.)		**	***	**
Pear (*Pyrus communis*)	**	**	*	**
Plums (*Prunus domestica*) **T**	**	***	*	*
Primrose (*Primula vulgaris*)		*	*	*
Snowy mespilus (*Amelanchier lamarckii*) **T**	**	*	*	*
Wallflower (*Erysimum* spp.) [*Erysimum* 'Bowles's Mauve']	**	*	**	**
Willow (*Salix* spp.) [*S. caprea*] **T**	***	***	***	***

Summer (*June, July, August*)

	Honey bees	Solitary bees	Long-tongued bumblebees	Short-tongued bumblebees
Agastache (*Agastache* spp.)	**	**	*	*
Astrantia (*Astrantia* spp.)	**	**	*	**
Barberry (*Berberis* spp.)	**	**	**	***
Bellflowers (*Campanula* spp.)	**	**	***	***
Borage (*Borago officinalis*)	***	***	***	**
Buddleia (*Buddleja* spp.) [*B. globosa*]	**	**	**	**
Calamint (*Calamintha nepeta*)	**	***	***	**
Cardoon (*Cynara cardunculus*)	***	**	**	**
Catmint (*Nepeta* spp.)	**	**	***	**
Comfrey (*Symphytum* spp.)	*	**	***	***
Cosmos (*Cosmos bipinnatus*)	**	**	**	**
Cotoneaster (*Cotoneaster* spp.) [*C. horizontalis*]	***	**	**	***
Cotton thistle (*Onopordum acanthium*)	**	**	***	***
Culver's root (*Veronicastrum virginicum*)	***	**	***	***
Dahlia (*Dahlia* spp.)	**	**	**	*
Echium (*Echium* spp.) [*E. pininana, E. wildpretii*]	***	***	***	***
Evening primrose (*Oenothera biennis*)	*	*	*	*
Firethorn (*Pyracantha coccinea*)	**	**	**	**
Foxgloves (*Digitalis* spp.)		*	***	***
Foxtail lily (*Eremurus* spp.)	**	**	**	**
Germanders (*Teucrium* spp.) [*Teucrium × lucidrys, T. hircanicum*]	**	*	**	**
Geums (*Geum* spp.)	**	*	**	**
Giant scabious (*Cephalaria gigantea*)	**	*	**	***
Globe thistle (*Echinops* spp.)	**	**	**	**
Hardy geranium (*Geranium* spp.) [*Geranium* 'Rozanne', *G. phaeum, G. pratense, G. molle*]	**	**	**	**
Honesty (*Lunaria annua*)	**	*	*	*
Hyssop (*Hyssopus officinalis*)	***	*	**	**
Knapweed & cornflower (*Centaurea* spp.)	**	***	***	***
Lamb's ear (*Stachys byzantina*)	**	**	***	**
Lavender (*Lavandula* spp.) [*Lavandula × intermedia* 'Gros Bleu']	***	**	***	***
Lime (*Tilia* spp.) **T**	***	*	**	***
Mexican sunflower (*Tithonia rotundifolia*)	**	**	*	**
Oregano (*Origanum vulgare*) [*O. vulgare* var. *album*]	**	**	***	**

Persicaria (*Persicaria* spp.)	**	*	**	**
Phacelia (*Phacelia tanacetifolia*)	***	*	*	**
Plume thistle (*Cirsium rivulare*)	**	**	**	**
Poached egg flower (*Limnanthes douglasii*)	**	*	*	**
Pot marigold (*Calendula officinalis*)				
Purple loostrife (*Lythrum salicaria*)	**	*	**	*
Rosemary (*Salvia rosmarinus*)	**	*	**	**
Salvias (*Salvia* spp.)	**	*	**	**
Scabious (*Scabiosa* spp.)	**	***	***	***
Sea holly (*Eryngium* spp.)	**	**	***	***
Seven son flower tree (*Heptacodium miconioides*) **T**	**	*	*	**
Siberian bugloss (*Brunnera, macrophylla*)	**	**	*	**
Snowberry & coralberry (*Symphoricarpos albus* and *S. orbiculatus*)	**	*	*	**
Society garlic (*Tulbaghia violacea*)	**	*	*	**
Spiked speedwell (*Veronica spicata*)	***	**	***	***
Sunflowers – annual & perennial (*Helianthus* spp.)	**	**	**	**
Sweet rocket (*Hesperis matronalis*)	**	**	*	*
Tansy (*Tanacetum vulgare*)	*	**	*	*
Thyme (*Thymus* spp.)	**	**	**	**
Toadflax (*Linaria* spp.) [*L. purpurea*]	*	*	**	*
Veronicas (*Veronica* spp.)	***	**	***	***
Viper's bugloss (*Echium vulgare*) [*E. vulgare* 'Blue Bedder']	***	**	***	***
Wild Privet (*Ligustrum vulgare*)	**	*	**	**

Autumn (*September, October, November*)

Aster (Michaelmas daisy) (*Aster/Symphyotrichum* spp.) [*Aster × frikartii, S. novi-belgii*]	***	***	**	**
Bee-Bee Tree (*Tetradium daniellii*) **T**	***	*	*	
Black-eyed Susan (*Rudbeckia* spp.)	**	**	**	**
Bugbane (*Actaea simplex*)	**	*	**	**
Chaste tree (*Vitex agnus-castus*)	**	*	**	**
Chinese gall (*Rhus chinensis*) **T**	**	*	*	**
Coneflower (*Echinacea* spp.)	**	**	**	**
Goldenrod (*Solidago* spp.)	**	**	**	*
Heleniums (*Helenium* spp.) [*H. autumnale, Helenium* 'Moerheim Beauty']	**	**	**	**

Ivy (*Hedera helix*)	***	**		*
Japanese anemone (*Anemone* spp.)	**		*	*
Japanese loquat (*Eriobobotrya japonica*) **T**	**	*	*	**
Japanese pagoda tree (*Styphnolobium japonicum*) **T**	*		*	*
Mignonette (*Reseda odorata*)	***	**	*	**
Sedum (stonecrop) (*Hylotelephium* spp.) [*H. spectabile*]	**	**	**	**
Verbena (*Verbena* spp.)	**	*	**	**

Winter (*December, January, February*)

Crocus (*Crocus* spp.)	***	**		*
Daphnes (*Daphne* spp.) [*D. bholua, D. mezereum*]	**			
Early stachyurus (*Stachyurus praecox*)	**	*	*	
Hazel (*Corylus avellana*) **T**	**			
Hellebores (*Helleborus* spp.)	*	*		*
Oregon grape (*Mahonia aquifolium*)	*	*		*
Snowdrops (*Galanthus* spp.)	**			
Sweet box (*Sarcoccoa confusa*)	**			
Viburnum (*Viburnum* spp.) [*Viburnum × bodnantense* 'Dawn']	**	*	*	**
Winter aconite (*Eranthis hyemalis*)	**			*
Winter clematis (*Clematis* spp.) [*C. cirrhosa*]	*			
Winter-flowering cherry (*Prunus × subhirtella*)	**	*	*	*
Winter heath (*Erica carnea*)	**	*	**	**
Winter honeysuckle (*Lonicera fragrantissima*)		*		*
Yellow monkswort (*Nonea lutea*)	*	*	**	*

Further reading

One of the great joys of having an interest in bees and beekeeping is the huge wealth of literature on the subject. To keep bees successfully you will need several practical reference books at least, after which, if you are interested, there are books about every conceivable related subject. There is also a huge number of historical books, dating back centuries. Many are readily available for surprisingly reasonable prices, while some might fetch tens of thousands at auction – in which case, pray for good honey harvests!

Aston, D. & Bucknall, S. *Plants and Honey Bees: Their Relationships.* (Northern Bee Books, 2004).

Benton, T. *Bumblebees.* (Collins, New Naturalist Library, 2006).

Benton, T. & Owens, N. *Solitary Bees* (Collins, New Naturalist Library, 2023).

Chapman, R. *How to Make a Warming Cabinet for Two Honey Buckets.* (BeeCraft, 2012).

Chikta, L. *The Mind of a Bee* (Princeton University Press, 2023)

Cramp, D. *A Practical manual of Beekeeping.* (Spring Hill, 2018).

Davis, C. F. *The Honey Bee Inside Out.* (BeeCraft, 2015).

Davis, C. F. *The Honey Bee Around and About.* (BeeCraft, 2014).

Davis, I. & Cullum-Kenyon, R. *The BBKA Guide to Beekeeping.* (Bloomsbury, 2015).

Falk, S. & Lewington, R. *Field Guide to the Bees of Great Britain and Ireland.* (Bloomsbury, 2015).

Gould, J. L. & Gould, C. *The Honey Bee.* (Scientific American Library, 1988).

Gregory, P. *Healthy Bees are Happy Bees.* (BeeCraft, 2018).

Kirk, W. D. J. & Howes, F. N. *Plants for Bees.* (International Bee Research Association, 2012).

Ollerton, J. *Pollinators and Pollination.* (Pelagic Publishing, 2021).

O'Toole, C. *Bees: A Natural History.* (Firefly Books, 2013).

Oudolf, P. & Gerritsen, H. *Planting the Natural Garden.* (Timber Press, 2019).

Owen, J. *Wildlife of a Garden: A Thirty-Year Study.* (Royal Horticultural Society, 2010).

Seeley, T. D. *Honeybee Democracy.* (Harvard University Press, 2010).

Seeley, T. D. *The Lives of Bees: The Untold Story of the Honey Bees in the Wild.* (Princeton University Press, 2019).

Tautz, J. *The Buzz About Bees.* (Springer-Verlag, 2008).

Waring, A. & Waring, C. *The Bee Manual.* (Haynes Publishing, 2011).

Winston, M. L. *The Biology of the Honey Bee.* (Harvard University Press, 1987).

An excellent source of new and second-hand bee books: www.northernbeebooks.co.uk

Magazines

BeeCraft (monthly) www.bee-craft.com

Websites

The British Beekeepers Assocation
www.BBKA.org.uk

The National Bee Unit/BeeBase:
www.nationalbeeunit.com

To report suspected Asian hornet sightings, email alertnonnative@ceh.ac.uk

LEFT You might have difficulty burying your head in a fuchsia. Try a good bee book instead.

Acknowledgements

To my wife, Varsha, who not only encourages, supports and enables me in everything I do but also tolerates cars full of bees, plans curtailed by swarms and an almost permanently sticky kitchen floor. If only I could say that her reward was a cupboard full of honey – but she can't stand the stuff. Big hugs to Charlie, Ruby and Lily, who have grown up thinking that walking through clouds of bees to get to the front door is normal.

Many thanks to Alice Ward, commissioning editor at Bloomsbury, who asked me to write this book and more or less left me to get on with it until encouragement was needed as deadlines loomed. Heather Bradbury oversaw the final stages of editing and design and made a hectic process delightfully pleasant. Three cheers for designer Susan McIntyre who somehow took too many words and photos and made them fit beautifully on the page – and rearranged them when I changed my mind.

My gratitude goes to Prof Ted Benton for help with species identification, Prof Jürgen Tautz for scientific advice, Dr George McGavin for entomological pointers, Jekka McVikar for help with herbs and Norman Carreck for checking in several places that I hadn't made a mess of things.

Thanks as well to Stephen Fleming, my co-editor on *BeeCraft* magazine, who shares my fascination with bees and much else, and with whom I have enjoyed various bee-related adventures that have fed one way or another into this book.

Finally, thank you to Emma Morley; a beekeeping friend who very obligingly, and sometimes at little notice, modelled for photos. Her colour-coordinated beesuit and hives are an art director's dreams come true.

I am very grateful to the following for their kind permission to use their photographs:

Amanda Berry; 261. Anita Burrough; 127B. Amanda Hayes; 40, 41. Animal and Plant Health Agency (APHA) Crown copyright; 142B, 141BL. Dan Basterfield; 153BL. Glen Harper, Plymouth Electron Microscopy Centre, University of Plymouth; SEM images 25–28, 107. National Bee Supplies; 41T, 52. Kirsty Stainton; 145BL. Panoramedia CC BY-SA 3.0; 179T. Roland Steinman CC BY-SA 2.0; 179B.

Index

Page numbers in *italic* indicate figures.

abdomen, honey bees 27
acarine 142
after-swarms *see* cast swarming
agricultural crops 288–91, *289*, *290*, *291*
agrochemicals 14–15, 16, 212, 276, 288
allotments 44–6
American foulbrood (AFB) 140, *140*
amoeba 142
anatomy, honey bees 25–8, *25*, *26*, *27*, *28*
annual meadows 247, *247*, 248–9, *248*, *249*
antennae, honey bees 26, *26*

backfilling 129
Bailey comb change 156–7, *157*
bald brood 137, *137*
bee banks 179, 204–6, *205*
bee bread 108
bee escapes 120
bee sheds 44
bee space 46, *51*, 52, 76
bee suits 54, *54*
BeeCraft 20
beefly, dark-edged 206, *206*
beehives *see* hives
beekeeping
    on allotments 44–6
    beekeeping year 158–61
    dividing colonies 83–4, 87–90, *88*, *89*, 103
    feeding bees 129–31, *130*
    finding and marking queens 77–8, *77*, *78*
    in gardens 39–40, 43–4

guidelines 20
history of 17–18, 37
hive inspections 71–6, *72*, *73*, *74*, *76*, 85–6, 141
how many bees to keep 40
hygiene 135
impacts on wild bees 16–17, 18–19
learning about 20–1
local bees 59–60
moving colonies 99–102, *100*, *101*, *102*
natural beekeeping 37–8
neighbours and 39–40, 43, 67
nucleus colonies 60–2, *60*, *62*
numbers of colonies in UK 16, 18, 19
record-keeping 78–9
replacing comb 153–7, *153*, *154*, *155*, *156*, *157*
responsible 18–20
seasons 21, 158–61
in sheds 44
smoking 68, 69–71, *70*
stings 40, 66–7, *67*
swarm collecting 63–6, *63*, *64*, *65*, 68
uniting colonies 103–4, *103*, *104*, *105*, 132
winter preparations 131–2, *131*, *132*
see also equipment; health issues; hives; pests; re-queening; swarm management
bees
    declines 15–17
    evolution 9–10
    importance of 12–13
    threats to 13–17
beeswax 126–7, *126*, *127*
blowtorches 56

brace comb 46, 52, 76, *76*, 83
breeding programmes 17–18, 59
British Beekeepers Association (BBKA) 16
brood boxes 46, 48
brood development 32–4, *32*, *33*, *34*
brood food 26, 27, 32, 34, 108
brood frames *see* frames
brood nests 73, 75–6, *75*
Buckfast bees 59
bumblebee species
    brown-banded carder 170, 175
    buff-tailed 166, 171, *171*, *178*, 216, 222, 276
    Burbut's cuckoo 173
    common carder 164, 175, *175*, 179, *179*, 216, 282
    early 168, 174, *174*
    field cuckoo 175
    forest cuckoo 174, 177, *218*
    garden 173, *173*, 216
    gypsy cuckoo 172
    moss carder 175
    red-shanked carder 175
    red-tailed 169, 176, *176*
    red-tailed cuckoo 176
    shrill carder 15–16, *15*, 175
    southern cuckoo 167, 171
    tree 16, 177, *177*, 179, *179*, 268
    white-tailed 172, *172*
bumblebees 10, 165–79, 214
    colony size 167
    crop pollination 276
    cuckoo bumblebees 167–8, *167*
    declines 15–17
    identification 168–77
    impacts of beekeeping on 16–17, 18–19
    life cycle 165–7

nesting sites 165–6, *166*, 178–9, *178*, *179*, 268–9
    threats to 13–17
    tongue length 216
    in winter 222, *223*
burr comb 126

cappings wax *126*, 127
Carr, William Broughton 47
Carson, Rachel 14
cast swarming 82, 91, *91*
chalkbrood 136, *136*
chilled brood 137, *138*
chimneying 117
chronic bee paralysis virus (CBPV) 142, *142*
clearer boards 120, *120*
climate change 15, 21, 222, 261, 293
clipping queens 86, *86*
clothing, protective 54–5, *54*
clusters, winter 132–3, *133*
cob bricks 206–7, *207*
Colletes bees
    Davies' colletes 187, *187*
    ivy bee 16, 187, *187*
colonies 29–31, *29*, *30*, *31*
    dividing 83–4, 87–90, *88*, *89*, 103
    moving 99–102, *100*, *101*, *102*
    nucleus colonies 60–2, *60*, *62*
    uniting 103–4, *103*, *104*, *105*, 132
    see also hives; re-queening; swarming
colony collapse disorder (CCD) 16, 142–3
comb replacement 153–7, *153*, *154*, *155*, *156*, *157*
comb wax 126
comfrey 280–1, *280*
compost heaps 179
conservation areas 19
crops
    agricultural 288–91, *289*, *290*, *291*
    comfrey 280–1, *280*
    fruit crops 277, *277*
    green manures 280, *280*
    herbs 281–6, *281*, *282*, *283*, *284*, *285*, *286*, *287*

pollination 12–13, 275–7
    vegetable plots 278–81, *278*, *279*, *280*
cross-pollination 277
crownboards *46*, 49
crystallisation of honey 124–5, *124*
cuckoo bumblebees 167–8, *167*
    Burbut's 173
    field 175
    forest 174, 177, *218*
    gypsy 172
    red-tailed 176
    southern *167*, 171
cuckoo solitary bees 10, 183, 195, *195*, 201–2
cut comb honey 123–4

damp gardens 260, 261–3, *261*, *262*, *263*
dance, waggle 82, 113–14, *113*, *114*
Darwinian beekeeping 37
deformed wing virus (DWV) 144, *144*
diapause 183
die-offs 16
diseases *see* health issues
domestication 17–18
double-walled hive 47
drone brood monitoring 145, *145*
drone congregation areas (DCAs) 35, *35*
drone-laying queens 138, *138*
drones 31, *31*
    death 31, 35
    development 32–3, *33*, *34*
    mating 35, *35*
    swarming and 84, 85, *85*
dummy boards 50

egg laying
    bumblebees 166
    honey bees 30, 32, *32*, 76, 83, 132, *133*, 138, *138*
    solitary bees 182–3
emergency feeding 131
emergency queen cells 90, 93–4, *93*, 96, 97, 98
endophallus 35
equipment 54–7
    feeders 130

frame rests 57, *57*
gloves 54–5, *54*
hive tools 55, *55*
honey extractors 121–2, *122*, *123*
protective clothing 54–5, *54*
queen cages 56, *56*
queen catchers 56, *56*
smokers 55, *55*, 68, 69–71, *70*
strainers 123, *123*
uncapping tools 56, *56*, 121, *121*
washing soda crystals 57
wax extractors 126, *126*
    see also hives
European foulbrood (EFB) 139, *139*, 153
extractors
    honey 121–2, *122*, *123*
    wax 126, *126*
extrafloral nectaries 111, *111*, 215, 216
eyes, honey bees 26, *26*

feeding bees 129–31, *130*
feet, honey bees 28, *28*
flatpack hives 50–3, *51*, *52*
Flow hive 47–8
flower bees 10, 190, 195
    green-eyed 260
    hairy-footed 11, 180, 182, 190, *190*, 206, 216, 280
    potter flower bee 190
flower structure 214, *215*
flows, spring and summer 110–11, 117–19
fondant *130*, 131
forage
    finding 113–14
    pollen 107–8, *107*, 133, 214
    propolis 108–9, *109*
    water 109–10, *109*, 133
    see also nectar
foragers 31, 99–101, 107, 114, *115*, *116*
forest honey 111
foundation *51*, 52–3, 83
frame rests 57, *57*
frames 48, *49*
    assembling 50–3, *51*, *52*
    extracted 124, *124*

foundation *51*, 52–3, 83
inspecting 73–6, *74*, *75*
warm way or cold way 53, *53*
Frisch, Karl von 9, 113
fruit crops 277, *277*
furrow bees 188, 203, *204*
common 188, *188*
orange-legged 189, *189*
sharp-collared *181*

gardens 211–13
bee banks 179, 204–6, *205*
biodiversity in 211
bumblebee nesting sites 165–6, *166*,
178–9, *178*, *179*, 268–9
compost heaps 179
damp gardens 260, 261–3, *261*, *262*,
263
flowering lawns 243–6, *244*, *245*,
246
herbs 281–6, *281*, *282*, *283*, *284*, *285*,
286, *287*
keeping bees in 39–40, 43–4
meadows 242, 247–50, *247*, *248*,
249, *250*, *251*
organic gardening 212
ponds 256, 257, *258*, *259*
solitary bee nesting sites 184, 203–
7, *203*, *204*, *205*, *206*, *207*, 269
sustainability 212
vegetable plots 278–81, *278*, *279*,
280
wild gardens 252–4, *252*, *253*, *254*,
255
see also nest boxes for solitary bees;
planting for bees
glands 31
hypopharyngeal 27, 32, 114
mandibular 26, 32
Nasonov 27
wax 27, 31
gloves 54–5, *54*
green manures 280, *280*

habitat loss 13–14
haemolymph 27
hairs, honey bees 26, 28

hamuli 27, *27*
head, honey bees 26–7, *26*
health issues 135–43
acarine 142
American foulbrood (AFB) 140,
140
amoeba 142
bald brood 137, *137*
brood inspections 141
chalkbrood 136, *136*
chilled brood 137, *138*
chronic bee paralysis virus (CBPV)
142, *142*
colony collapse disorder (CCD) 16,
142–3
comb replacement 153–7, *153*, *154*,
155, *156*, *157*
deformed wing virus (DWV) 144,
144
drone-laying queens 138, *138*
European foulbrood (EFB) 139,
139, 153
hygiene 135
laying workers 138, *138*
nosema 141–2, *141*
poisoning 142, 143
quarantine 135
sacbrood 137, *137*
signs of healthy brood 136, *136*
see also pests; varroa mites
heather honey 111, 118–19
herbicides 15, 212
herbs 281–6, *281*, *282*, *283*, *284*, *285*,
286, *287*
hive tools 55, *55*
hives
assembling 50–3, *51*, *52*
bee escapes 120
bee space 46, *51*, *52*, 76
brace comb 46, 52, 76, *76*, 83
brood boxes 46, 48
choosing site for 43–6
clearer boards 120, *120*
comb replacement 153–7, *153*, *154*,
155, *156*, *157*
construction materials 49–50
crownboards 46, 49

dummy boards 50
floors 46, 48
foundation *51*, 52–3, 83
frames 48, 49, 50–3, *51*, *52*, *53*
hygiene 135
inspections 71–6, *72*, *73*, *74*, *76*,
85–6, 141
insulation 131–2, *131*
moving 99–102, *100*, *101*, *102*
for natural beekeeping 37
nuc boxes 60, 61, 62, 64–5, *64*
orientation 45
queen excluders 46, 48–9, *49*, 116
roofs 46, 49
runners 52
skeps 37, *38*, 63, *63*
spacing between 45
stands 46, 49
supers 46, 49, 83, 116–17, *117*
types 37, 39, 46–8, *46*, *47*
varroa monitoring 145–6, *145*, *146*
warm way or cold way 53, *53*
winter preparations 131–2, *131*,
132
Hoffman frames *51*, 52
homing instinct 99–101
honey 8, *8*, 114–26
adding supers for 49, 116–17, *117*
average harvest 124
crystallisation 124–5, *124*
cut comb honey 123–4
extraction 120–4, *121*, *122*, *123*, *124*
harvesting 117, 119–24, *119*, *120*
honeydew honey 111
how bees make honey 114–15, *115*,
116, *117*
jarring 123, 125, *125*
labelling regulations 126
nectar sources 110–11, *110*, *111*, *112*,
117–19, *118*, *119*
selling 125–6
spring honey 117
storing 123, 124–5
straining 123, *123*
summer honey 117–19
warming 125, *125*
water content 119

honey bees
anatomy 25–8, *25, 26, 27, 28*
breeding programmes 17–18, 59
brood development 32–4, *32, 33, 34*
colonies 29–31, *29, 30, 31*
declines 16
die-offs 16
domestication 17–18
egg laying 30, 32, *32,* 76, 83, 132, 133
evolution 9–10
homing instinct 99–101
honey production 114–15, *115, 116, 117*
life cycle 32–5, *32, 33, 34*
mating 35, *35*
as native species 17
pheromones 26, 27, 28, 30, 35, 64–5, *65*
species and subspecies 9–10, *9, 10,* 59, *60*
threats to 16
tongue length 216
waggle dance 82, 113–14, *113, 114*
winter behaviour 132–3, *133*
see also beekeeping; forage; health issues; pests; swarming
honey flows 110–11, 117–19
honey stomach 27
honeydew 111, *111,* 215, *215*
honeydew honey 111
horizontal top-bar hive 37, *39*
hornets
Asian 132, 151–2, *151, 152*
European 149, *152*
house bees 31, 114, *116*
hygiene 135
hypopharyngeal glands 27, 32, 114

inner-city beekeeping *17, 18, 19*
insecticides 14–15, 16, 212, 276, 288
inspection tray monitoring 145, *145*
insulation 131–2, *131*
Integrated Pest Management (IPM) 146–8, *146, 147, 148*
intensively farmed areas 19
ivy bee 16, 187, *187*

June gap 111, 117

Krefeld Society 13

Langstroth, Lorenzo 46
Langstroth hive 46, *46*
larvae
bumblebees 166–7
honey bees 32–3, *33, 34, 34,* 81, 85–6, 136, *136*
solitary bees 183
lawns
flowering 243–6, *244, 245, 246*
solitary bee nesting sites 203–4, *203, 204*
leafcutter bees 193, *193,* 195, 198, 201–2
patchwork *183,* 193, *193*
Willughby's 193, *193*
woodcarving 288
legs, honey bees 28, *28*
local bees 59–60

magnifying glasses 57
mandibles, honey bees 26, *26*
mandibular glands 26, 32
marking queens 77, *77, 78, 78*
mason bees 191, 195, *197*
blue 192, *192*
red *181,* 191, *191,* 200, *200, 201,* 276, *277, 286*
red-tailed 192, *192*
spined 221
mating
bumblebees 167
honey bees 35, *35*
solitary bees 182, *182*
meadows 242, 247–50, *247, 248, 249, 250, 251*
mice 149, *149*
mining bees 184, *206*
ashy 185, *185,* 203, *203*
chocolate 186, *186*
common mini-miner 187, *187*
Maidstone 183
orange-tailed 186, *186*
sandpit *204*

tawny 185, *185,* 203, *203*
movable-frame hive 37

Nasonov glands 27
Nasonov pheromone 27, 64–5, *65*
National Bee Unit (NBU) 135
National hive *46, 47, 47,* 48
natural beekeeping 37–8
nectar 110, 214–15
collection 114, *115*
extrafloral 111, *111,* 215, *216*
honeydew 111, *111,* 215, *215*
important UK plants for 110–11, *110, 112,* 117, 118–19, *118, 119*
processing into honey 114–15, *116*
spring and summer flows 110–11, 117–19
nectaries
extrafloral 111, *111,* 215, *216*
in flowers 110, 214–15, *215*
neighbours 39–40, 43, 67
neonicotinoid insecticides (neonics) 14–15, 16, 276
nest boxes for solitary bees 196–203, *196, 197*
boxes 197–8
grooved boxes 200, *200*
maintenance 202–3
nesting blocks *197,* 198–9, *198*
nesting tubes *196,* 199–200, *199*
predators and parasites 201–2, *202*
siting 200, *201*
nesting sites
bumblebees 165–6, *166,* 178–9, *178, 179,* 268–9
solitary bees 184, 203–7, *203, 204, 205, 206, 207,* 269
nosema 141–2, *141*
nuc boxes 60, 61, 62, 64–5, *64*
nucleus colonies 60–2, *60, 62*
nucleus method of swarm management 87–90, *88, 89,* 103

organic gardening 212
orientation flights 62
ovaries 27, 32

ovarioles 32
Owen, Jennifer 211

parasites 201–2, *202*, 206, *206*
    see also varroa mites
parthenogenesis 32
perennial meadows 242, 247–8, 250,
    250, *251*
pests
    Asian hornets 132, 151–2, *151*, *152*
    mice 149, *149*
    small hive beetles 151, *151*
    Tropilaelaps mites 151, *151*
    wasps 148–9, *148*, *149*
    wax moths 136, *136*, 150, *150*
    woodpeckers 150, *150*
pets 67
pheromones 26, 27, 28, 30, 35, 64–5, *65*
piping 82
planting for bees 219–21
    100 best garden plants 293–5
    annuals 231, 232, *232*, *233*
    biennials *230*, 231, *231*, 234, *234*, *235*
    bulbs 226–8, *226*, *227*, *228*, *229*
    climbers 268
    damp gardens 260, 261–3, *261*, *262*,
        263
    flowering lawns 243–6, *244*, *245*,
        246
    flowering time 220–1
    hedges 265–6, *265*, *266*
    herbaceous perennials 236–9, *237*,
        238, *239*, *240–1*
    herbs 281–6, *281*, *282*, *283*, *284*, *285*,
        286, *287*
    meadows 242, 247–50, *247*, *248*,
        249, *250*, *251*
    origin 219–20
    ponds 256, *257*, *258*, *259*
    quantity 220
    shape, size and colour 219
    shrubs 263, *263*, 267–8, *267*
    trees 263, *263*, 268–72, *269*, *270*, *271*,
        272, *273*
    wild gardens 252–4, *253*, *254*, *255*
    winter and early spring flowers
        222–4, *222*, *223*, *224*, *225*

plants and bees 213–16
    extrafloral nectaries 111, *111*, 215, *216*
    flower structure 214, *215*
    honeydew 111, *111*, 215, *215*
    how flowers attract bees 216, *217*
    pollen 107–8, *107*, 133, 214
    pollination 12–13, 213–14, *214*,
        275–7
    propolis 108–9, *109*
    top UK plants for honey
        production 110–11, *110*, *112*,
        117, 118–19, *118*, *119*
    see also crops; nectar
plasterer bees
    Davies' colletes 187, *187*
    ivy bee 16, 187, *187*
    yellow-faced bees 182, 189, *189*,
        196, 198
play cups *see* queen cups
poisoning *142*, 143
pollen 107–8, *107*, 133, 214
pollen baskets 28, 214
pollen brushes
    honey bees 24, 28, *28*
    solitary bees 182, *183*, *213*, 214
pollen press 28
pollination 12–13, 213–14, *214*, 275–7
polystyrene hives 47, 50
ponds 256, *257*, *258*, *259*
Porter bee escapes 120
proboscis, honey bees 26–7
propolis 46, 52, 108–9, *109*, 127
protective clothing 54–5, *54*
pupae 33, *33*, 34, *34*

quarantine 135
queen cages 56, *56*
queen catchers 56, *56*
queen cells
    emergency 90, 93–4, *93*, 96, 97, 98
    sealed 90–1
    supersedure 94, *94*
    swarm 34, 81, 82, *83*, 84–6, 88–9,
        88, 90–1, 93, *93*
queen cups 34, 81, 84–5, *85*, 86, 93
queen excluders 46, 48, 49, 116
queen substance 30

queen trapping 147
queens 30, *30*, 92
    clipping 86, *86*
    development 32, 33–4, *34*, 81, 82,
        85–6, 90
    drone-laying 138, *138*
    egg laying 30, 32, *32*, 76, 83, 132,
        133, 138, *138*
    finding and marking 77–8, *77*, *78*
    mating 35, *35*
    in nucleus colonies 60–1
    ovaries 27, 32
    pheromones 28, 30, 35
    piping 82
    stings 82
    see also re-queening; swarm
        management; swarming

radial extractors 122, *122*
record-keeping 78–9
refractometers 119
reproduction 32–5, *32*, *33*, *34*, *35*
re-queening 94–9
    after losing a swarm 90–1
    colony split method of queen
        raising 95–7
    queen introduction 97–8, *99*
    queen-rearing 95
resin bees *197*, 198
royal jelly 32, 34, 81, 85
rural areas 19

sacbrood 137, *137*
sacrificial drone comb 146–7, *146*
scissor bees *197*, 238
scout bees 82, 113–14, *113*, *114*
sealed queen cells 90–1
seasons 21, 158–61
sexual organs 27, 35
sharp-tailed bees *195*, 201–2
sheds, keeping bees in 44
shook swarm technique 147, 153–6,
    154, *155*, *156*
skeps 37, *38*, 63, *63*
small hive beetles 151, *151*
smokers and smoking 55, *55*, 68,
    69–71, *70*

solitary bees 10, 181–207
    cavity-nesting species 191–5
    crop pollination 276
    cuckoo solitary bees 10, 183, 195, *195*, 201–2
    declines 15–17
    ground-nesting species 184–90
    identification 183–95
    impacts of beekeeping on 16–17, 18–19
    life cycle 181–3, *181, 182*
    nesting sites 184, 203–7, *203, 204, 205, 206, 207*, 269
    numbers of species 183
    pollen collection *182, 183, 213, 214*
    threats to 13–17
    tongue length 216
    see also nest boxes for solitary bees
spermatheca 32, 35
spiracles 27
sternites 27
stings
    anatomy 27, 31
    getting stung 40, 66–7, *67*
    queens 82
strainers 123, *123*
suburban areas 19
sugar-shake varroa monitoring 146, *146*
sun 45
supers *46*, 49, 83, 116–17, *117*
supersedure queen cells 94, *94*
swarm cells *see* swarm queen cells
swarm collecting 63–6, *63, 64, 65, 68*
swarm management 34, 82–91
    adding supers 83
    clipping queens 86, *86*

dividing colonies 83–4, 87–90, *88, 89*
    inspections 85–6
    nucleus method 87–90, *88, 89*, 103
    sealed queen cells 90–1
    signs of swarming 84–5, *84, 85*
swarm queen cells 34, 81, 82, 83, 84–6, 88–9, *88*, 90–1, 93, *93*
swarming 29, 33–4, 40, *80*, 81–2, 87, *91*
    cast swarming 82, 91, *91*
    factors affecting 82–4
    shook swarming 153–6, *154, 155, 156*
    signs of 84–5, *84, 85*

tangential extractors 122, *122*
tergites 27
thorax, honey bees 27
tongue length 216
tracheae 27
trophallaxis 26–7, *116*
Tropilaelaps mites 151, *151*

uncapping 121, *121*
uncapping tools 56, *56*, 121, *121*
urban beekeeping *17*, 18, 19

varroa mites 16, 61, 136, 143–8, *143*
    controlling 132, 146–8, *146, 147, 148*
    life cycle 144
    monitoring for 145–6, *145, 146*
    symptoms of 144–5, *144, 145*
vegetable plots 278–81, *278, 279, 280*

waggle dance 82, 113–14, *113, 114*
walls and bricks, solitary bee nesting sites 206–7, *206, 207*

washing soda crystals 57
wasps 148–9, *148, 149*
water
    collection and use by bees 109–10, *109*, 133
    water content of honey 119
wax glands 27, 31
wax mirrors 27
wax moths 136, *136*, 150, *150*
WBC hive 47, *47*
wild gardens 252–4, *252, 253, 254, 255*
wind 45
wings, honey bees 27, *27*
winter
    bumblebees in 222, *223*
    feeding bees 129–31, *130*
    honey bee behaviour 132–3, *133*
    moving hives 102
    preparing for 131–2, *131, 132*
    varroa treatment 132, 147–8
woodpeckers 150, *150*
wool carder bee 182, *182*, 194, *194*
worker food 32
workers 30–1, *31*
    death 31
    development 32–3, *33, 34*
    foragers 31, 99–101, 107, 114, *115, 116*
    house bees 31, 114, *116*
    laying by 138, *138*
    scout bees 82, 113–14, *113, 114*
    winter bees 31, 132

yellow food 32
yellow-faced bees 182, 189, *189, 196*, 198